AF615551

# Diffusion Processes

M. H. Jacobs

Springer-Verlag Berlin Heidelberg New York 1967

Professor Dr. M. H. JACOBS
University of Pennsylvania
School of Medicine
Department of Physiology
Philadelphia, Pa. USA

Reprint from
Ergebnisse der Biologie
Zwölfter Band, 1935



Library of Congress Catalog Card Number AC 38 - 3748
Printed in Germany
Title No. 1440

## Preface

A basic tenet of present day biophysics is that flows in biological systems are causally related to forces. A large and growing fraction of membrane biophysics is devoted to an exploration of the quantitative relationship between forces and flows in order to understand both the nature of biological membranes and the processes that take place on and in these membranes. This is why the discussion of the nature of diffusion is so important in any formal development of membrane biophysics. This was equally true twenty years ago when tracers were just beginning to be used for the measurement of membrane processes. We turned naturally to the great treatises on the physics of diffusion and the flow of heat where, to be sure, we could dig out the information that was needed. It was a great joy then to come across this masterful and scholarly discussion on diffusion written for biologists of a physical turn of mind by MERKEL JACOBS. Here were to be found not only the equations that were basic to our knowledge, but also a careful, accurate and logical explanation, both of the physical principles and the mathematical steps. It soon became apparent that we could not keep that one volume of Ergebnisse der Biologie on indefinite loan from the library, and we then found, by good fortune, a remaindered copy of this particular issue. It has become a well-thumbed and treasured possession of the laboratory. I am very glad that others now will have a similar opportunity to learn about diffusion from MERKEL JACOBS and that this volume will be widely available again.

Subsequently I had occasion to tell JACOBS how much pleasure this article had given us and asked him how he had found the time to do all the work. He had broken his leg and had written it to while away the time the when he was confined to the hospital. This story seems very illustrative of JACOBS' devotion to science. In our laboratory which has been concerned for many years with the permeability of red blood cells, we have always had great respect for JACOBS and for his insight into the forces that cause flows across that specific membrane. We go back again and again to his papers and to his methods, particularly the mathematical ones, and always unearth new things of value — an observation as true today as when we first began research on red cells. That we are not alone in our high regard for JACOBS is illustrated by another anecdote. I once asked USSING how he had come to make his theoretical examination of the difference between osmotic flow and the diffusion of water. USSING's interest had been aroused by the clear statement of the problem and the suggestions of JACOBS in an article reviewing "cell permeability with particular reference to the erythrocyte".

April 17, 1967 A. K. SOLOMON

Biophysical Laboratory
Harvard Medical School
Boston, Mass. USA

# Diffusion Processes.

By **M. H. JACOBS**, Philadelphia.

## Table of Contents.

Page

## 1. Introduction.

The subject of diffusion is one of great practical and theoretical importance in the biological sciences. Every cell, of every organism, at every moment of its existence, is dependent upon this process for supplying it with necessary materials from its surroundings, for distributing these and other materials within its boundaries, and for removing to a safe distance metabolic products which if allowed to accumulate would be injurious. However slightly different cells may resemble one another in other respects, they all show a common dependence upon this, the most widespread of all cellular activities.

Visible evidence of the universal importance of diffusion processes is furnished by many details of cytological structure — in particular, by the small size of the ultimate physiological units of organisms, by which such processes are facilitated, and by the universal presence in these units of differentially permeable membranes, by which the same processes are limited and controlled. In all parts of the bodies of the higher organisms special structural adaptations associated with diffusion are the rule rather than the exception. As examples, chosen almost at random, may be mentioned the peculiar shape of the mammalian erythrocyte (HARTRIDGE 1920, PONDER 1925, 1926); the spacing of the capillaries in the tissues (KROGH 1919b); the thin walls and enlarged surfaces of organs of respiration in general, whether lungs, gills or insect tracheae; the expansion of absorptive surfaces by structures as unrelated as the intestinal villi of mammals and the leaves and root-hairs of plants — in short, almost no organ of importance in either animals or plants fails to betray by at least some feature of its structure an intimate relation to diffusion processes.

Though the practical aspects of diffusion are the ones most usually emphasized by biologists, the process is also one of peculiar theoretical and even philosophical interest. Diffusion is one of the chief means by which, in accordance with the second law of thermodynamics, the distribution of matter and energy in the universe tends constantly to become less and less orderly and more and more of the sort that would result from the operation of the laws of chance. From this point of view, even the simplest organism is an almost incredibly improbable accumulation of matter, which might be expected when it changes at all to change in the direction of increasing probability of arrangement. Progress in the latter direction is, indeed, the rule after death, just as it is in the inorganic world. But in living organisms, though individual diffusion processes are found to occur in the expected manner, the sum-total of all such processes is typically in the opposite direction. Not only does the highly improbable arrangement of materials found in the fertilized ovum give place to enormously less probable ones as the development of the individual proceeds, but living matter, in general, seems in the course of centuries and of geological epochs constantly, and with only rare exceptions, to assume forms which from the standpoint of the distribution of matter are likewise less and less probable. It is somewhat paradoxical that individual development and the evolution of the race should alike be so utterly dependent upon diffusion processes, which in their fundamental nature are diametrically opposed to both. The manner in which living organisms have succeeded in harnessing, so to speak, these essentially destructive processes and in utilizing them for constructive purposes is, in fact, one of the major mysteries of Biology.

Historically, there has always been a close relation between the study of diffusion and the biological sciences. The earliest experiments in this field (NOLLET 1748, FISCHER 1822, MAGNUS 1827, POISSON 1827, DUTROCHET 1827, JERICHAU 1835, BRÜCKE 1843, VIERORDT 1848, LUDWIG 1849, JOLLY 1849, etc.) had to do chiefly with the diffusion of water and solutes across animal membranes, and in most cases they admittedly had as their object the explanation of certain physiological processes in plants and animals. These early experiments with membranes prepared the way for the epoch-making work on osmotic phenomena of the plant-physiologist PFEFFER (1877) which was destined to have such far-reaching effects, not merely in plant and animal physiology, but in physical chemistry as well. In the meantime, the same early studies of diffusion across membranes undoubtedly stimulated the work on "free diffusion" of GRAHAM (1850a, b; 1851a, b, 1861, 1862a, b) and of FICK (1855), with which the modern history of diffusion processes may be said to have begun.

Another very important point of contact between the biological sciences and the subject of diffusion processes grew out of the observation of an English botanist, ROBERT BROWN (1828), that certain granules of vegetable origin under the microscope exhibit a continuous irregular movement which has since received his name. After a long history, whose details must be omitted for lack of space, BROWNian movement was definitely shown by EINSTEIN (1905), v. SMOLUCHOUSKI (1906), PERRIN (1909), SVEDBERG (1912) and others to be a visible manifestation of the mechanism of diffusion. In recent years, biologists have again become interested in BROWNian movement, this time as a useful tool for the investigation of some of the fundamental physical properties of protoplasm; references to some of this work will be given below (p. 111).

Because of the long and intimate relation between diffusion processes and the biological sciences, it is not inappropriate that a discussion of this subject should appear in the „Ergebnisse der Biologie". It should be noted that excellent treatises on diffusion and the closely related subject of heat conduction, written primarily from the point of view of the physicist, are already in existence (for example, BYERLY[1] 1893, INGERSOLL and ZOBEL 1913, CARSLAW 1921, FÜRTH 1927c, 1931a, etc.), but as far as the author is aware none of these has been prepared with the peculiar needs, and, in particular, with the limited acquaintance with higher mathematics of the average biologist in mind. The present review is therefore designed to fill an existing gap in the literature. Though it may itself at first sight appear unduly mathematical for a biological journal, it will be found on closer examination to contain nothing not readily intelligible to anyone who has mastered the general principles of the differential and integral calculus. In this respect, and in its frank selection of material because of its biological rather than of its mathematical or physical interest, it differs from its predecessors, for which it is in no sense intended to be a substitute, and to which, it is hoped, it may in many cases serve as an introduction.

To workers in the so-called exact sciences it may perhaps appear rash and even presumptuous for a biologist to attempt, with so little mathematical equipment, to deal with diffusion processes in living systems. Such systems are enormously more complicated than the relatively simple artificial ones usually studied by physicists, whose mathematical resources they frequently tax to the limit. It is, however, the very fact that living systems are so complex that justifies the type of treatment here adopted. The justification is, in fact, a twofold one. In the first place, it is utterly hopeless for the biologist with the means at present at his disposal, to reduce the variables that enter into his problems to the small number usually encountered in physical investigations. He is compelled, therefore, regretfully but of necessity, to be content with a lesser degree of precision in his results than that attainable in the so-called "exact sciences". It follows that in dealing with most biological problems it is not only useless, but actually unscientific, to carry mathematical refinements beyond a certain point, just as it would be both useless and unscientific to employ an analytical balance of the highest precision for obtaining the growth-curve of a rat.

In the second place, the field of biology, comprising as it does so many millions of species of plants and animals, among which are to be found the utmost conceivable differences in structure and activity, is

[1] The author is indebted to this excellent work for important parts of the mathematical treatment of the subjects dealt with in sections 6, 10 and 12.

so vast that the biologist is still in the position of an explorer in a newly discovered continent. His first task is to map out more or less roughly the main topographical features of the country — its rivers, lakes and mountains — after which accurate geodetic surveys may profitably be undertaken. In discovering a new mountain range the possession of accurate surveyor's instruments is not only of little assistance but may even be a handicap. Remembering, therefore, that the biologist is still for the most part an explorer rather than a surveyor, and that what he needs most at present — to continue the figure of speech — is not complicated instruments so much as an ax, a rifle and a compass, this simplified mathematical treatment of the subject of diffusion processes is presented without further apologies.

## 2. Diffusion and probability.

Experience has shown that whenever local concentration differences are found to exist in an otherwise uniform body of solution, sufficiently large to permit its study by ordinary chemical methods, these differences tend with time to become less and less pronounced, and finally to disappear. This spontaneous process, which in a homogeneous system must ultimately bring about uniformity of concentration everywhere within the system, is called diffusion, and the final state of the system one of diffusion equilibrium. Experience shows further that when diffusion equilibrium has been attained, local concentration differences of appreciable magnitude never again appear in the system without the expenditure of energy from external sources. Diffusion as commonly observed is, therefore, a typical one-sided, irreversible process illustrating in a visible form the second law of thermodynamics.

If we inquire more closely why diffusion always takes place in certain types of systems and why, after once having occurred, the process is irreversible, we are led to an explanation which is essentially mathematical rather than physical. There is, in fact, no purely physical reason why diffusion *must* proceed in the commonly observed direction and why the process can *never* under any circumstances undergo a spontaneous reversal. The question is merely one of mathematical probability, with the odds so overwhelmingly in favor of the occurrence of the process in the usual manner that nobody has ever seen, or could reasonably expect to see, in a region of considerable size, any departure from the so-called diffusion laws.

In the case of very minute systems, however, in which the number of molecules involved is small, the case is different. If such a system could be studied in detail, it would be found that equalization of concentration in it does not proceed smoothly towards a final permanent equilibrium, but rather that it occurs irregularly, with frequent reversals of direction, and that the final state is characterized, not by permanence, but by continual fluctuations about a purely statistical equilibrium position. Cases of this sort are of some theoretical interest to the biologist, since the possibility has been seriously suggested that deviations from the usual so-called "laws" of diffusion in very minute regions of a living cell might conceivably have appreciable physiological consequences (FREUNDLICH 1919, DONNAN 1927). Strictly speaking, of course, there is no fixed size above which large systems become different from small systems. What is true of the latter is true of the former except for a difference in the magnitude of the deviations that may be expected to occur.

The general question of the statistical basis of diffusion processes is of such fundamental scientific and even philosophical importance that it may be profitable to consider it in a somewhat more concrete manner. Imagine a very small spherical or other symmetrical region filled with water and containing at first only 10 solute molecules. It is obvious that in such a system we could never with any certainty predict the positions of these molecules. At room temperature they would, if of ordinary size, move with an average velocity of the order of magnitude of 100 meters per second or more. In an aqueous medium, in every second, each one would collide with countless millions of water molecules, and with each collision the direction and velocity of its movement would be altered. The path of an individual molecule under such conditions is utterly unpredictable; it is in fact an almost perfect example of the operation of the laws of pure "chance"; and the laws of "chance" applied to one or a few molecules can give very little useful information. As the number of units dealt with increases, however, these same laws increase in value, until finally, with the unimaginably large numbers of molecules that enter into ordinary diffusion processes, they lead to a degree of certainty scarcely exceeded in any other phase of human experience.

Returning to the simple case of 10 molecules, we have to do with a very small region and with velocities comparable to those of a rifle bullet. We may, therefore, assume that any effect of an initial distribution would so quickly be obliterated that at the time of an imaginary first observation the probability that a given molecule would be found in, say, the upper half of the region would be exactly $\frac{1}{2}$, and the same that it would be found in the lower half. It follows that the probability of finding all the molecules in one specified half of the system at any chosen instant would be $\left(\frac{1}{2}\right)^{10}$ or 1 in 1024, and the likelihood of finding any other number can readily be calculated by the theory of permutations and combinations. Assuming that no distinction is to be made between individual molecules, the number of chances in $2^n$ of finding a number, $r$, when the total number is $n$ is given by the formula

$$\frac{\underline{|n}}{\underline{|r}\;\underline{|n-r}}\,.$$

Applying this formula and tabulating the results, we have for the case in question

| Number | Number of chances in 1024 |
|---|---|
| 10 | 1 |
| 9 | 10 |
| 8 | 45 |
| 7 | 120 |
| 6 | 210 |
| 5 | 252 |
| 4 | 210 |
| 3 | 120 |
| 2 | 45 |
| 1 | 10 |
| 0 | 1 |

Even in this simple system we might begin to see, though in a very crude way, some indication of the laws of diffusion. If, for example, at some given time we happened to observe, say, 9 molecules in the upper and one in the lower half of the region in question, we should probably find at the next observation, assuming the passage of sufficient time to permit thorough „mixing", that a decrease in the upper and an increase in the lower half had occurred, i. e., that a tendency towards an equalization of concentration in the two regions had been manifested. There is no physical reason why a change in this direction must occur; but with only one chance in 1024 of a change in the opposite direction, and only 10 in 1024 of no change at all, the probability that we should observe "diffusion" of material from a region of higher to one of the lower concentration even in this simple hypothetical case is very great indeed; we might, in fact, repeat such an observation a considerable number of times without observing an exception to what we might, therefore, be tempted to speak of as a "law".

There is no fundamental distinction between the hypothetical case just described and systems of the sort that come under actual scientific observation; the only difference is one of probability. It is instructive to make certain further calculations in connection with a kind of system that might actually be studied in practice. Suppose, for example, that 1 cc. of a molar solution of dextrose be introduced without mixing below 1 cc. of water in a cylindrical vessel. We wish to consider the subsequent behavior of such a system from the standpoint of probability alone, purposely neglecting the factor of the time required for diffusion, which has no effect on the direction and the position of equilibrium of the process, and which could in any event be regulated at will by altering the diameter of the vessel. The number of solute molecules is so large (i. e., $6 \times 10^{20}$) that the improbability that they could ever again spontaneously congregate in one half of the solution after an approximate equality of distribution had been attained by mixing or by diffusion becomes, practically speaking, an impossibility. The mathematical chance of their doing so amounts, to be exact, to $\left(\frac{1}{2}\right)^{6 \times 10^{20}}$. While, therefore, by human agency we may readily start with an arrangement of molecules of the sort mentioned, we should not expect that by diffusion processes alone it could ever be restored after the two halves of the solution had in any way become mixed.

Of more interest is the fact that even a barely detectable increase of concentration in one half of the vessel after equilibrium has been attained is likewise so improbable as, practically speaking, to be impossible. Noting that in the case of 10 molecules the probabilities of the different possible arrangements are all given by the successive terms of the binomial expansion $\left(\frac{1}{2}+\frac{1}{2}\right)^{10}$, the same relation can be used, in principle, for a very large number by taking advantage of the fact that in the binomial expansion $\left(\frac{1}{2}+\frac{1}{2}\right)^{n}$, as $n$ increases indefinitely, the coefficients of the successive terms approach more and more nearly to the values of a series of equidistant ordinates of the normal probability curve

$$y=\frac{1}{\sigma\sqrt{2\pi}}e^{-\frac{x^2}{2\sigma^2}}$$

where $\sigma$, the so-called standard deviation of the statistician, must be properly determined to fit the case in question. Since it may readily be seen from the tabulated values of the probability integral that random variations greater than $3\sigma$ occur only about three times in a thousand, it follows that greater variations than this would scarcely ever be encountered in an ordinary series of observations.

Now it is shown in all the standard works on statistical methods that if $n$ events be considered, and it the probability of a success (in this case the presence of a molecule in a selected half of the solution) be represented by $p$ and of a failure (i. e., its absence in the same region) by $q$, then

$$\sigma=\sqrt{npq}.$$

In the present case $p=q=\frac{1}{2}$, and $n=6\times 10^{20}$, so $\sigma$ is equal to $1.22\times 10^{10}$ and $3\sigma$ to approximately $3.7\times 10^{10}$. Variations greater than this, i. e., greater than 0.000,000,012% of the theoretical equilibrium value of $3\times 10^{20}$, which would be utterly inappreciable by any available methods of chemical analysis, could scarcely be expected to occur.

While it is true that we should only rarely expect variations of more than $3\sigma$, there is a remote possibility of much larger ones. How remote is the possibility of encountering a variation great enough to be detected by chemical methods? Suppose that by some method of analysis we could measure a difference of concentration of 1 part in 100,000 or 0.001 per cent from the equilibrium concentration. In the present case this would mean the detection of a deviation of $3\times 10^{15}$ molecules from the mean equilibrium number of $3\times 10^{20}$ per cc. This deviation amounts roughly to 250,000 times $\sigma$. The probability that by chance alone a variation could exceed 250,000 $\sigma$, while almost infinitesimally small, can nevertheless be calculated so long as the normal probability distribution may be assumed to hold. Strictly speaking, this distribution implies an infinite rather than a very large though finite number of molecules, but except for much larger deviations than the one in question it is permissible to employ the usual formula for such cases, namely,

$$P = \frac{e^{-z^2}}{z\sqrt{\pi}}\left(1 - \frac{1}{2z^2} + \frac{1.3}{(2z^2)^2} - \ldots\right)$$

where $z$ is the deviation divided by $\sigma\sqrt{2}$. Applying this formula to a deviation of 250,000 $\sigma$, $P$ proves to be approximately 1 divided by a number of the order of magnitude of 1 followed by 13,000,000,000 zeros — a number which if printed in the type used on this page would cover a distance of approximately 23,000 kilometers.

Calculations such as these explain perhaps better than is possible in mere words the fundamental nature of diffusion processes — the reason why such processes occur, the reason why they approach a definite predictable equilibrium and the reason why this equilibrium when once reached seems to be maintained indefinitely without further change. They also serve to emphasize the important fact, already mentioned, that the laws of diffusion, in their last analysis, are based upon mathematical rather than upon purely physical principles.

## 3. FICK's law.

The inevitability of the transfer of material from one region to another in homogeneous systems showing concentration gradients is obvious from the principles discussed in the preceding section. The manner in which this transfer takes place may next be examined. The modern theoretical treatment of diffusion may be said to have begun with the clear recognition by FICK (1855) that this process is analogous in most respects to the conduction of heat in solids, which had already been treated mathematically by FOURIER. The same idea had much earlier been expressed in rather general terms by BERTHOLLET (1803), but FICK was the first to give it real definiteness, and in particular, to put it to the test of experiment. The fundamental assumption of FICK was that the rate of diffusion across any plane at right angles to the direction of diffusion bears a simple linear relation, which may be quantitatively defined by a constant, called by him the diffusion constant, to the concentration gradient across the plane in question.

Stated in mathematical terms, FICK's law is

$$dQ = -DA\frac{\partial u}{\partial x}dt \qquad (1)$$

where $dQ$ represents the amount of material diffusing in the time $dt$, during which all conditions may be considered to remain constant, across a plane of area $A$ at right angles to the direction of diffusion, the concentration gradient at the plane being $\frac{\partial u}{\partial x}$. Throughout this paper, following many earlier authors, concentration will be represented by the symbol $u$ rather than $c$, since the latter letter is so frequently employed as a constant, and since in order to apply diffusion equations to the flow of heat, and vice-versa, it is desirable to use a terminology which fits either case equally well. The symbol $D$ appearing in equation (1) was represented by Fick by $k$ and was assumed by him to be a constant for all values of $u$, though experimental work soon showed that this assumption is justified at best only as a somewhat rough approximation; for this reason the term diffusion coefficient is preferable, and will hereafter be used in this paper. $D$ evidently represents the amount of material that in unit time and with unit concentration gradient would cross a plane of unit area at right angles to the direction of diffusion.

The unit of concentration, $u$, may be defined as one unit of quantity in unit volume. For the latter, the cubic centimeter is used rather than the liter, and if, as is generally the case, concentrations in a given problem are originally expressed in mols per liter they must first be divided by 1000 before being introduced into Fick's equation. Since the unit in which $Q$ is measured, whatever it may be, also enters into the definition of the unit of concentration, the same numerical value of $D$ must obviously apply to all cases of diffusion regardless of whether measurements are made in terms of mols, grams, number of individual molecules, etc. per cc. It will be observed that $D$ has the dimensions $\frac{\text{cm}^2}{\text{unit of time}}$, since on applying Fick's equation to any experimentally determined data, we have the general relation

$$D = \frac{a \text{ units of quantity}}{(b\ \text{cm}^2)\ (c \text{ units of time}) \left(\frac{d \text{ units of quantity}}{e\ \text{cm}^3}\right) \Big/ f\ \text{cm}} = \frac{a e f}{b c d}\ \text{cm}^2 \text{ per unit of time.}$$

The unit of time employed in connection with diffusion coefficients is sometimes the second, less frequently the minute, and most commonly the day, because of the slowness of diffusion in vessels of the sizes generally used for determinations of $D$. A final important point about Fick's equation is the negative sign appearing before $D$, which in published discussions of diffusion problems is occasionally, by oversight, omitted. Such an omission is serious if further mathematical use of the equation is to be made, since diffusion necessarily occurs in the direction of decreasing rather than of increasing concentrations, and a positive sign before $D$ would indicate the reverse to be the case.

At first sight, FICK's law of a direct linear relation between the rate of diffusion and the concentration gradient appears not unreasonable; but it is, in fact, only the form assumed for infinitely dilute solutions by a much more complicated law or series of laws. It bears the same relation to the more general law that BOYLE's law for perfect gases does to those governing actual gases, or that VAN'T HOFF's law of osmotic pressure does to those actually obeyed by concentrated solutions. This similarity between the laws of FICK, BOYLE and VAN'T HOFF is more than a superficial one, since the three are in reality related to one another, and similar factors may be responsible for deviations from all of them.

The theoretical basis for FICK's law, as well as the reasons for its limited applicability, are perhaps best brought out by a method of treatment used by NERNST (1888) and later by EINSTEIN (1908, 1922). The latter author will here be followed fairly closely except for the use of a somewhat different definition of the term "osmotic pressure". Imagine a case of diffusion in one dimension parallel to the long axis of a vessel or tube, which for simplicity may be considered to have a cross section of unity. The effects of gravitation may be neglected, and uniformity of temperature and of all conditions other than concentration may be assumed to exist. Select any layer, taken across the vessel at right angles to the direction of diffusion, and of infinitesimal thickness, $dx$. This layer, having unit cross section, will also have the volume $dx$. Let the concentration of solute within it be $u$ mols per cc; the number of solute molecules it contains will therefore be $uNdx$, where $N$ is the number of molecules in one mol.

Now suppose that each free surface of the layer is readily permeable to water, but completely impermeable to the solute. By the known properties of osmotic systems, there must be a movement of water through such a layer from the more dilute to the more concentrated solution, and if external forces be absent, the layer itself will be moved in the opposite direction until conditions on its two sides have become completely equalized. It is possible, however, to stop the flow of water and therefore the movement of the layer at any time by applying an appropriate external pressure to the more concentrated of the two adjacent solutions. Osmotic pressure may be defined (see LEWIS 1923, WASHBURN 1921, etc.) as the pressure which must be applied to a given solution under some given conditions to make the escaping tendency of the solvent which it contains equal to that of the pure solvent under the same conditions, and the osmotic pressure as thus defined may be taken as equal to $p$ for the solution in immediate contact with the layer on the one side and to $p + dp$ for that on the other, the corresponding values of $x$ being $x$ and $x + dx$. Under these conditions, water will tend to pass in the positive direction and in consequence to move the layer in the negative direction, the effective pressure available for this purpose being the difference between the two osmotic pressures. Since the area of the layer is unity, pressure and force will be numerically equal.

Now the total force exerted on the layer may be imagined to be applied to the transport of the number of molecules, $u N d x$, contained within it. The driving force, $k$, acting on a single molecule in the positive direction is therefore $-\frac{d p}{u N d x}$. Such a force would transport the molecule with some velocity, $v$, equal to $k/f$, where $f$ is the frictional resistence encountered by the molecule at unit velocity, which for the present need not be further defined than by the relation just stated. Substituting the value of $k$ given above, we have

$$v = -\frac{1}{u N f} \frac{d p}{d x}.$$

Now in a vessel of unit cross-section the rate of transport of material in mols per unit of time, i.e., $dQ/dt$, will be equal to $uv$. Furthermore, if the solution be assumed to be very dilute, or to obey the same laws as dilute solutions, then by VAN'T HOFF's Law, $p = u R T$ and the equation becomes

$$\frac{dQ}{dt} = -\frac{R T}{N f} \frac{d u}{d x}. \tag{2}$$

Equation (2) is seen to be the same as FICK's equation (1) for diffusion across unit area except that $D$ is now replaced by $RT/Nf$. This relation is of great importance, particularly when the magnitude of $f$ can be determined experimentally or can be calculated. The latter possibility frequently exists, since according to STOKES' law, for a sphere of radius $r$ (which must not be too small in comparison with that of the surrounding molecules) and a viscosity $\eta$ of the liquid medium

$$f = 6 \pi \eta r.$$

For any case to which STOKES' law is applicable we therefore have

$$D = \frac{R T}{N} \cdot \frac{1}{6 \pi \eta r}. \tag{3}$$

This important equation is commonly known as the EINSTEIN or the STOKES-EINSTEIN equation (EINSTEIN 1905), though it was obtained independently somewhat earlier by SUTHERLAND (1904, 1905). A similar equation differing only in the presence of a numerical factor, $\frac{64}{27}$, was also derived by a different method about the same time by v. SMOLUCHOWSKI (1906); but, according to LANGEVIN (1908) who repeated v. SMOLUCHOWSKI's calculations, this factor ought not to appear, and in any case the SUTHERLAND-EINSTEIN equation is in much better agreement with the known facts, and was used by v. SMOLUCHOWSKI himself in his later work.

The SUTHERLAND-EINSTEIN equation has very often been used to obtain information about the radii of the diffusing molecules or colloidal particles from the observed values of their diffusion coefficients. A typical case of this sort, of physiological interest, has recently been discussed by NORTHROP and ANSON (1929). By a method described below (p. 66) they obtained experimentally a value of $D$ for hemoglobin of 0.0420 $cm^2$/day. The other quantities appearing in equation (3) with the exception of $r$ are known; the following are the values used for purposes of calculation: $R = 8.3 \times 10^7$, $N = 6.06 \times 10^{23}$, $T = 278$, $\eta = 0.01519$. The value of $r$ as thus calculated proves to be $2.73 \times 10^{-7}$ cm. Assuming that the hemoglobin molecule is spherical and that its specific gravity, $g$, is the same as that of crystalline hemoglobin, i.e., 1.33, the molecular weight calculated by the formula

$$m = \frac{4}{3} \pi r^3 g N$$

proves to be $68{,}500 \pm 1{,}000$.

Similar applications of the law in question have been made with varying degrees of success by other workers. In many cases, even where it might not be expected to hold, the law seems to give good results, but in general it must be used critically and with due regard for complicating factors not taken into account in its derivation. Thus, both theory and experience indicate that it is strictly applicable to single molecules only if these are fairly large in comparison with the molecules of the solvent. In the case of most ordinary substances there seems to be a closer inverse proportionality between $D$ and the square root of the molecular weight than between $D$ and the radius of the molecule, which might be expected in turn to be roughly proportional to the cube root of the molecular weight. In fact, the relation, $D\sqrt{MW} = K$ has been much used in diffusion studies, though it shows many exceptions, particularly when the substances compared are not closely related chemically. The question of the relation of $D$ to the molecular weight and the radius of the diffusing molecule is discussed among others by RIECKE (1890), EULER (1897), THOVERT (1902, 1910), ÖHOLM (1910, 1912b), ZEILE (1933), etc.

A further relation deducible from the SUTHERLAND-EINSTEIN equation is that for the same substance in different media or at different temperatures the product of the diffusion coefficient by the viscosity of the medium ought to be constant. While this relation has been found to hold very satisfactorily in many cases, THOVERT (1904, 1914), ÖHOLM (1912c), COHEN and BRUINS (1923b), it is by no means universal (ÖHOLM 1913, MILLER 1924) and it must therefore be applied with caution. In general, however, the relation is an important one and seems to be chiefly though by no means wholly responsible for the values assumed by the temperature coefficients of diffusion processes in aqueous media, which are roughly of the order of magnitude of those associated with the viscosity of water (for fuller data see ÖHOLM 1912b).

A very important point connected with the use of equation (3), sufficiently obvious at present, but until recently largely overlooked, is that it is not permissible to apply this equation to the calculation of the radii of diffusing particles in colloidal systems if these particles are ions and are accompanied by other ions of much smaller size. As will be shown below, the observed diffusion coefficient of an electrolyte depends upon the behavior of both of its ions, and it follows that the readily measured rate of diffusion of, for example, the colored ion of an organic dye-salt is, in itself, of little use as a basis for calculating of the radius of the dye molecule. Indeed, Bruins (1931a) and others have recently obtained with certain colloids diffusion coefficients which would seem absurdly great, exceeding as they do those of many crystalloids, were the presence and the contribution to the observed effect of the "Gegenionen" neglected. Since this important principle was not taken into account by most of the earlier workers, a large proportion of the published applications of the Sutherland-Einstein equation to questions of molecular and particle size in colloidal systems is of doubtful value. For further discussion of this point see Bruins (1931a, b, c, 1932), Hartley and Robinson (1931), Samec, Knop and Panovič (1932), McBain and Dawson (1934), and McBain, Dawson and Barker (1934).

The question of the exact behavior of electrolytes in diffusion processes, particularly when their concentrations are high, or when several electrolytes diffuse simultaneously, is much too complex to fall within the scope of the present paper. A general introduction to this question is given by Nernst (1926), and many further details will be found in the following incomplete list of references to the original literature: Nernst (1888), Arrhenius (1892), Wiedeburg (1892, 1899), Behn (1897), Bose (1899), Abegg and Bose (1899), Thovert (1902b), Haskell (1908), Osborne and Jackson (1914), v. Hevesy (1913a, b), Walpole (1915), Goldschmidt (1929), Hartley (1931), McBain and collaborators (1931—1934), Sitte (1932), Onsager and Fuoss (1932), Davies (1933).

Only one question concerns us here, namely, the fact that an electrolyte, though dissociating into two or more ions, behaves in a manner that can be described by a single diffusion coefficient, which, in the case of sufficiently dilute solutions, may be calculated with a very fair degree of accuracy by means of the so-called Nernst equation. The original treatment of this problem by Nernst (1888) for the case of a uni-univalent electrolyte may be somewhat simplified in a manner very similar to that already followed above for a non-electrolyte (see Nernst 1926 and Taylor 1924). For a more general treatment of the behavior of electrolytes whose ions have any valences, the discussion by A. A. Noyes quoted by Haskell (1908) may be consulted.

It is known from studies of the electrical conductance of solutions that different ions have different mobilities, which depend upon their masses, degrees of hydration, etc. Let the mobilities of the two oppositely charged ions of a uni-univalent electrolyte of concentration $u$ be $U$ and $V$, respectively. Because these mobilities are, in general, different, one ion will tend to diffuse more readily than the other, but the electrostatic forces so set up will prevent any complete separation of the two sorts of ions, and the net effect will be that the rate of diffusion of the more slowly moving ion will be accelerated by that of its more rapid partner, and vice versa, giving a single rate of diffusion, represented by an appropriate diffusion coefficient, for the salt as a whole.

Suppose now that diffusion is occurring parallel to the long axis of a vessel of uniform cross-section, which for simplicity may be taken as unity. Consider as before (p. 11) a movable elementary layer having the thickness $dx$, lying at right angles to the direction of diffusion, and let the concentrations and osmotic pressures on the two sides of this layer be $u$ and $p$, and $u + du$ and $p + dp$, respectively. The volume of the layer is $dx$ and the amount of salt which it contains is $udx$ mols; this amount is supposed to be completely dissociated. By applying exactly the same principle as that already used in deriving the EINSTEIN equation for non-electrolytes, it is seen that the force acting on one mol of salt in the positive direction is $-\frac{1}{u} \cdot \frac{dp}{dx}$. This force, if the two kinds of ions were free from each other's influence, would therefore tend to produce movements in the positive direction in the time $dt$ of

$$-Uu\left(\frac{1}{u}\frac{dp}{dx}\right)dt \text{ and } -Vu\left(\frac{1}{u}\frac{dp}{dx}\right)dt.$$

Such movements, however, would set up electrostatic forces which would act in the same direction as the osmotic forces for the one ion and in the opposite direction for the other. Representing in the proper units the electrostatic potential by $E$, its effect on the movements of the two ions would be, respectively,

$$-Uu\frac{dE}{dx}dt \text{ and } +Vu\frac{dE}{dx}dt.$$

With equal rates of diffusion of the two ions it follows that

$$-U\left(\frac{dp}{dx} + u\frac{dE}{dx}\right)dt = -V\left(\frac{dp}{dx} - u\frac{dE}{dx}\right)dt = dQ.$$

By the elimination of $\frac{dE}{dx}$ and the substitution of $p = uRT$, since the solution is assumed to be very dilute

$$dQ = -\frac{2UV}{U+V}RT\frac{du}{dx}dt. \tag{4}$$

On comparing this equation with FICK's law, it follows that

$$D = \frac{2UV}{U+V}RT. \tag{5}$$

This simple and important relation has been shown to hold with a very satisfactory degree of accuracy for dilute solutions of electrolytes.

See in this connection ÖHOLM (1905) and among more recent workers ULLMANN (1927), FÜRTH and ULLMANN (1927) and ZUBER and SITTE (1932).

Equations (2) and (4) not only furnish a theoretical basis for FICK's law, but they serve at the same time to indicate its limitations. For example, the assumption made in deriving both equations that $p = uRT$ is known not to be strictly correct for any actual solution of finite concentration; while for solutions of high concentration it ceases to be even approximately true. The osmotic laws applicable to concentrated solutions, particularly to those of electrolytes, are in fact complex and vary from substance to substance. It would be unreasonable, therefore, to expect a constancy of $D$ in cases where the relation of osmotic pressure to concentration is itself subject to change with concentration.

Furthermore, the frictional resistance, $f$, in equation (2) and the mobilities of the two ions in equation (4) which may be considered to be constant in sufficiently dilute solutions are by no means so under other conditions. In the case of non-electrolyte solutions, attractive forces between the individual solute molecules are assumed to be absent in infinitely dilute solutions, just as they are in the case of an imaginary "perfect gas", but with increasing concentration they begin to appear, and may lead in extreme cases to "association" and polymerization, with changes in the radii of the diffusing molecules, and in any event to effects which prevent the individual molecules from diffusing entirely freely as the theory demands that they should. At the same time, changes in "hydration" may also occur, which in some cases may take the form of fairly definite changes in the effective radii of the diffusing molecules, and in others may be limited to a less sharply defined influence on the surrounding water molecules, which nevertheless tends to affect the rate of diffusion.

Effects of both sorts are particularly striking in aqueous solutions of strong electrolytes in which the electrostatic forces of repulsion and attraction between ions of the same and of different sign, respectively, and between the ions and the surrounding water dipoles are so complex, even in solutions of moderate concentration, that it has only been in recent years that attempts to deal mathematically with such systems have been even partially successful. The most cursory examination of recent papers dealing with applications of the DEBYE-HÜCKEL theory of strong electrolytes to problems of diffusion (ONSAGER and FUOSS 1932, SITTE 1932, etc.) will show the unlikelihood, not merely of finding constant values of $D$ for an electrolyte over any considerable range of concentration, but even of finding values of $D$ which change with concentration according to any simple law.

From time to time efforts have been made to determine the laws of diffusion that apply to systems in which it is impossible to treat

an infinite system the distribution of material at any time in layers I to IV of the actual system. It is obvious from the purely random nature of molecular movements that in an infinite system half of the diffusing molecules that reach the upper boundary of layer IV will go forward and the other half backward; in the closed system just discussed, all must go backward; in the system now under consideration none will go backward. Evidently, therefore, by keeping the outer boundary constantly at the concentration zero, a backwardly directed stream exactly like the forwardly directed one which can be observed in layers 5, 6, 7, etc., of an infinite system is, in effect, removed. It follows that to find the amounts of material in layers I to IV inclusive it is necessary to subtract rather than to add the amounts calculated for layers 5 to 12 of an infinite system.

The situation is exactly as if a negative stream proceeded into the body of the solution from its boundary, equal except in sign to the positive stream which in an infinite system moves in the opposite direction. This negative stream might, for clearness, be thought of as composed of actual molecules having a negative sign, which algebraically could neutralize an equal number of positive molecules. Such a stream would evidently be reflected at the bottom of the vessel, but at the top of the vessel, which is open, not only would it not be reflected, but even half of it would fail to return as in an infinite system. By exactly the same principle as before, therefore, it would be necessary to subtract the negative stream that would theoretically pass from the vessel into an undisturbed infinite column of water; this is evidently equal except in sign to the positive stream that in the original infinite system would pass from layer 12 to layer 13. But since the effect of subtracting a negative stream is the same as adding a positive stream, we should have finally:

| | |
|---|---|
| IV | 4 — 5 — 12 + 13 |
| III | 3 — 6 — 11 + 14 |
| II | 2 — 7 — 10 + 15 |
| I | 1 — 8 — 9 etc. |

Another example of the principle of independent diffusion streams of considerable physiological interest is the following: Suppose, to take a concrete case, that a muscle originally possessing a uniform internal concentration, $a$, of lactic acid be placed in an unstirred solution entirely free from this substance. Diffusion of the lactic acid from the muscle to the solution will occur in a decidedly complex manner, which will be discussed below, and need not be considered in detail at this point. Suppose now that a similar muscle also having an internal concentration of $a$ be placed in another unstirred solution whose volume is equal to that of the first solution, but whose initial lactic acid concentration is $b$ (where $a > b$). In view of the known complexity of diffusion processes, it may perhaps not be immediately obvious that the amounts of lactic acid that will escape from the two muscles in any given time, must always bear to each other the simple numerical relation $a/(a-b)$. That this is the case, however, provided that FICK's law applies, follows at once from the principle of independent diffusion streams.

The molecules inside the second muscle may be divided into two imaginary groups, one having a concentration of $b$ and the other a concentration of $a-b$. The first group will be in exact equilibrium with the molecules of the same substance outside the cell, and while individual molecules can be exchanged between the two groups, no transfer of material between them can lead to visible concentration differences. There is left, therefore, only the unbalanced concentration, $a-b$, within the cell to be considered. If the relation between the two effective concentrations be: $a = n(a-b)$, we may think of diffusion in the first muscle as composed of $n$ independent streams each exactly like the single stream in the second muscle. It makes no difference therefore how complex the actual process of diffusion may be in each case; under otherwise equal conditions the losses of lactic acid from the two muscles must always be in the simple proportion $a/(a-b)$.

This principle has recently been put to practical use in determining the actual concentration within a muscle of diffusible creatine and urea (EGGLETON 1930) and phosphates (SEMEONOFF 1931). For such cases, two other methods have for a long time been available. The first is to allow diffusion to occur between the muscle and a known external solution until equilibrium is practically reached, and then from the amount of material transferred and the quantity of water in the muscle to estimate the concentration originally existing in the latter. This method requires an inconvenient amount of time, and may also permit unknown and uncontrollable changes to occur within the muscle during the course of the experiment. The other older method is to find by trial a solution in which neither gain nor loss of material occurs. This method has the advantage of involving only a short exposure of the muscle to the solution, but the disadvantage of sometimes requiring a considerable number of solutions and, even worse, a considerable number of muscles. The method of EGGLETON has the double advantage of permitting the desired concentration to be determined in a relatively brief time, and with the employment of only two external solutions and a pair of symmetrical muscles from the same animal. The precautions necessary in using the method are merely to make the comparison between two muscles that are as similar as possible, and to keep the external volumes and all other conditions equal in the two cases.

The theory of the method is as follows: Let the initial concentration of the diffusing substance in the muscle be $C_M$. Suppose that in the first experiment the external concentration change from $C_1$ to a final average value of $C_2$. The amount of material that has entered the muscle must, therefore, be $(C_1 - C_2) V$, the assumption of course being made that $V$ remains constant. Represent the same quantities in the second experiment in which the time is the same as before, by $C'_1$, $C'_2$ and $(C'_1 - C'_2) V$. By the principle of independent diffusion streams, a part of the external concentration in each case may be thought of exactly balancing the initial internal concentration $C_M$, while a second part whose concentration is $C_1 - C_M$ and $C'_1 - C_M$, respectively, diffuses into the muscle as if no other solute were present. If the numerical value of $C_1 - C_M$ be $n$ times that of $C'_1 - C_M$ we can think of diffusion in the former case as composed of $n$ separate streams, each due to an independent concentration of $C'_1 - C_M$ and the amount diffusing in time $t$ must consequently be $n$ times as great for $C_1 - C_M$ as for $C'_1 - C_M$. We may therefore write

$$\frac{Q}{Q'} = \frac{(C_1 - C_2)\, V}{(C'_1 - C'_2)\, V} = \frac{C_1 - C_M}{C'_1 - C_M}$$

which by a suitable transposition gives

$$C_M = \frac{C_1 C'_2 - C_2 C'_1}{(C_1 - C_2) - (C'_1 - C'_2)}\,. \tag{6}$$

This is the same as EGGLETON's equation except for somewhat different symbols and for a difference in sign caused by the fact that in the present case $C_1$ and $C_2$ are both taken as greater than $C_M$ while EGGLETON considered $C_1$ to be greater and $C_2$ less than $C_M$.

As an example of the use of this method, the following typical case may be mentioned. In one of EGGLETON's experiments, a symmetrical pair of sartorius muscles from the frog were placed in two equal quantities of RINGER's solution containing different amounts of creatine. The concentration of creatine in the first solution, measured in mg./100 g. was then found to decrease from 108 to 98, while during the same time the concentration in the second solution increased from 34 to 39. Applying equation (6), we have for the calculated concentration within the muscle

$$C_M = \frac{108 \times 39 - 98 \times 34}{(108 - 98) - (34 - 39)} = 59\ \text{mg./100 g.}$$

## 5. The general diffusion equation; initial and boundary conditions.

FICK's equation (p. 9), while a concise mathematical statement of his law of diffusion, is not, as it stands, directly applicable to the solution of most diffusion problems. Among other disadvantages, it contains four variables, $Q$, $u$, $x$, and $t$. The number of these variables can, however, be reduced to three, and a partial differential equation can be obtained whose solution for many particular problems is easy.

This latter equation, sometimes known as FICK's second equation, and here referred to as the general diffusion equation, is fundamental for the mathematical treatment of diffusion problems, and will now be derived. Though most of the cases of diffusion later to be considered involve only one dimension, it will be advantageous first to obtain the more general three-dimensional equation, which may then be modified to fit any special simpler cases as they arise.

In any general case of diffusion that obeys FICK's law, consider the behavior of a rectangular element of the region involved, with infinitesimal sides $dx$, $dy$ and $dz$, and having at its center the concentration $u$. However complicated the process of diffusion may be, it can for the element in question be resolved into three streams at right angles to each other and parallel to the three coordinate axes. By FICK's law, the rate of flow into and out of the element will be proportional to the concentration gradients associated with the three diffusion streams; for simplicity it will be assumed that the proportionality can be expressed by a single constant, $D$, for all concentrations. Under these conditions the rate of flow across the middle of the element in the $x$-direction will be $-D\frac{\partial u}{\partial x}\,dy\,dz$.

For the inflow into the element in the same direction, the same law will be followed, but the gradient will be slightly different. Since the gradient is a function of $x$, its rate of change with $x$ will be $\frac{\partial}{\partial x}\left(\frac{\partial u}{\partial x}\right)$ and through the infinitesimal distance $\frac{dx}{2}$ within which this rate of change must, as a limiting condition, be constant, the change will be $\frac{\partial^2 u}{\partial x^2}\cdot\frac{dx}{2}$. The rate of inflow, $\frac{dQ}{dt}$, will therefore be $-D\left(\frac{\partial u}{\partial x}-\frac{1}{2}\frac{\partial^2 u}{\partial x^2}dx\right)dy\,dz$. Similarly the rate of outflow will be $-D\left(\frac{\partial u}{\partial x}+\frac{1}{2}\frac{\partial^2 u}{\partial x^2}dx\right)dy\,dz$. Subtracting the latter rate from the former we have for the rate of accumulation associated with the stream in the $x$-direction: $D\frac{\partial^2 u}{\partial x^2}\,dx\,dy\,dz$. Going through the same procedure for the other two dimensions, we obtain similar expressions involving $y$ and $z$ in place of $x$; the sum of the three is the total rate of accumulation. Now the rate of accumulation may also be expressed in a different way. For an infinitesimal element, within which the concentration may be considered to have the single average value $u$, this rate will be equal to the rate of change of concentration multiplied by the volume, i. e., to $\frac{\partial u}{\partial t}\,dx\,dy\,dz$. Equating the two expressions for the rate of accumulation, and cancelling out the common factor $dx\,dy\,dz$, we obtain the general diffusion equation for a constant diffusion coefficient, namely,

$$\frac{\partial u}{\partial t}=D\left(\frac{\partial^2 u}{\partial x^2}+\frac{\partial^2 u}{\partial y^2}+\frac{\partial^2 u}{\partial z^2}\right). \qquad (7)$$

When diffusion is in one dimension only, this equation assumes the simpler form

$$\frac{\partial u}{\partial t} = D \frac{\partial^2 u}{\partial x^2}. \tag{8}$$

It will be noted that equation (8) contains only three variables. For the important cases of diffusion in cylinders and in spheres, which will be discussed below, it is necessary to retain all three dimensions, but the coordinates may advantageously be changed from the rectangular to the cylindrical and the spherical systems, respectively. The methods by which this change can be accomplished, which are found in the standard mathematical text books, need not be discussed here, but the resulting equations will be useful for reference and may therefore be presented without further comment. In the order mentioned they are

$$\frac{\partial u}{\partial t} = D\left(\frac{\partial^2 u}{\partial r^2} + \frac{1}{r}\frac{\partial u}{\partial r} + \frac{1}{r^2}\frac{\partial^2 u}{\partial \Phi^2} + \frac{\partial^2 u}{\partial z^2}\right). \tag{9}$$

$$\frac{\partial u}{\partial t} = \frac{D}{r^2}\left[\frac{\partial}{\partial r}\left(r^2\frac{\partial u}{\partial r}\right) + \frac{1}{\sin\theta}\frac{\partial}{\partial\theta}\left(\sin\theta\frac{\partial u}{\partial\theta}\right) + \frac{1}{\sin^2\theta}\frac{\partial^2 u}{\partial\Phi^2}\right]. \tag{10}$$

For the special cases of diffusion processes which are symmetrica about the axis of a cylinder or the center of a sphere, these equations assume much simpler forms which will be treated in some detail below (p. 114).

Most of the subsequent mathematical discussion of diffusion processes will be based upon equation (8), and it will therefore be profitable to consider some of its general properties. This equation is a concise mathematical statement of what all diffusion processes in one dimension, which are governed by Fick's law, possess in common. Translated into words, its meaning is that the rate of change of concentration at any level is proportional to the rate of change of the concentration gradient at that level. That this should be true is obvious from the fact that the rate of accumulation of diffusing material depends upon the difference between the rates of inflow and of outflow into an element of infinitesimal thickness, and this, in turn, depends upon the rate at which the concentration gradient changes. When expressed in this way in non-mathematical language, the information supplied by the equation may seem to be disappointingly meagre, and much too general for application to the solution of some particular problem. In a sense, this is true. It is a characteristic of partial differential equations that the information they convey is always of a very general nature, in itself far removed from the limitations imposed by actual systems. But by appropriate mathematical methods, what is general in the diffusion equation can be combined with the special information available about any given problem in such a way as to lead to a solution which completely fits this particular problem.

In general, in order to obtain a solution of a problem of diffusion in one dimension, three special pieces of information are necessary.

The first of these is a description of the initial distribution of the diffusing material in the system; this is the so-called *initial condition*. Since, by hypotheses, diffusion is restricted to one dimension, $x$, no concentration gradients can occur in any other dimension, and $u$ must therefore have the same value everywhere in the system for the same value of $x$. The initial condition may, therefore, always be represented by an equation of the type

$$u = f(x) \text{ when } t = 0 .$$

In addition to the initial condition, it is necessary in a case of diffusion in one dimension to have information about the properties of the two boundaries of the system that cross the axis of diffusion, which in all the cases here considered will be assumed to cross it at right angles. Since, by hypothesis, diffusion occurs in no other direction, and since, therefore, no other boundaries of the system can have any influence on its general progress, two so-called *boundary conditions* are all that are needed. The absence of diffusion in the directions represented by $y$ and $z$ may mean either that the system is closed or that it is infinite in these directions; in either case there can be no gain nor loss of material except through the boundaries mathematically represented by the planes $x = 0$ and $x = H$, $H$ here and elsewhere being used to represent the height or thickness of the system in the direction of diffusion.

Boundary conditions are of different sorts; one of the commonest takes the form of a statement as to the concentration of the diffusing substance that exists at a significant boundary of the system throughout the time under consideration. Two boundary conditions of this sort can be stated in the following manner

$$u = f_1(t) \text{ when } x = 0$$
$$u = f_2(t) \text{ when } x = H .$$

While $u$, as indicated in these equations, may theoretically vary in any manner with time at either or both boundaries, most of the cases that have so far found practical applications in physiology involve constant conditions only. The more difficult general case is treated in the text books on heat conduction, and arises in certain important physical problems, but it will be omitted here.

A second very important type of boundary condition is that in which no material crosses the boundary. This condition applies not only to the bottom of an ordinary vessel in which diffusion is occurring from below upwards, but to the upper surface of the liquid as well, unless the solute be a volatile substance which can escape from such a surface. In all cases where a boundary is closed, i. e., is impermeable to the solute, and FICK's law is true, equation (I) shows that an absence of diffusion and a gradient of zero must always go together. The mathematical representation of this type of boundary condition for a completely closed system is therefore.

$$\frac{\partial u}{\partial x} = 0 \text{ when } x = 0$$

$$\frac{\partial u}{\partial x} = 0 \text{ when } x = H.$$

The solution of problems of diffusion in one dimension in finite systems requires a knowledge of one initial and of two boundary conditions. In infinite systems in which no boundaries cross the axis of diffusion, or in semi-infinite systems in which only one such boundary crosses it, the necessary information takes a different form. In these cases, instead of information about the concentrations existing at certain boundaries, the information is merely given that such boundaries are non-existant.

Certain special cases may arise in which less than a complete solution is desired, and in which a knowledge of the initial condition is not necessary. One such case has to do with the distribution of material at equilibrium. If the system be closed, i. e., if $\frac{\partial u}{\partial x} = 0$ for $x = 0$ and for $x = H$, then, since material can neither enter nor leave the system, we know that whatever its initial distribution may have been, at equilibrium its concentration must everywhere be equal to the total amount originally present divided by the volume of the system. For such a case the final concentration obviously will be

$$\frac{1}{H}\int_0^H f(x)\,dx.$$

However, in order to find the concentration at equilibrium, it is not at all necessary that the form of $f(x)$ be known, provided that in any other way we can secure information about the total quantity of diffusible material present in the system. Similarly, if one boundary condition be $\frac{\partial u}{\partial x} = 0$ and the other be $u = c$, or if both be $u = c$, we know that at equilibrium $u$ must everywhere be equal to $c$ regardless of the initial distribution. A commonly encountered case of this sort is the one in which $c$ has the value of zero.

An interesting and very important case arises when we have the two boundary conditions $u = c_1$ when $x = 0$ and $u = c_2$ when $x = H$. In such cases, a true equilibrium is evidently impossible, since the concentration within the system cannot at the same time become equal to both $c_1$ and $c_2$. Under these conditions, what happens is the same thing that occurs when a reservoir containing water at one fixed level is connected with a second reservoir containing water at another fixed level: a steady and constant stream tends to become established between the two fixed regions. Such a condition may be called a steady state to distinguish it from a true equilibrium in which the level is everywhere the same. A steady state in diffusion resembles an equilibrium in that $\frac{\partial u}{\partial t} = 0$ for both; but whereas in the simple type of equilibrium here under discussion $\frac{\partial u}{\partial x} = 0$ as well, in a steady state this is not the case. Steady states are of such importance in the study of diffusion processes that they will be discussed in some detail below (p. 61); what concerns us here is merely that the character of such a state depends only on the two boundary conditions and not at all on the initial condition.

In considering boundary conditions, it is very important to remember that they apply to the region in which diffusion actually takes place and not to some adjacent medium or media. Thus, to cite a concrete case, suppose that diffusion is occurring between two stirred aqueous solutions of concentrations $c_1$ and $c_2$, respectively, through a region of thickness $H$, to which certain appropriate equations apply. The character of the diffusion process represented by these equations is obviously dependent upon the conditions existing at the planes $x = 0$ and $x = H$, which must be thought of as belonging to the region covered by the equations. Whether these conditions are or are not substantially identical with those an infinitesimal distance away in the two contiguous media depends entirely upon the nature of the system.

If the region separating the two external media be a capillary tube filled with water, and so small in diameter that convection currents are not set up in it by the stirring necessary to maintain constant external concentrations, then for practical purposes it is correct to use for such a system the boundary conditions $u = c_1$ when $x = 0$; $u = c_2$ when $x = H$. The same would be true of an artificial membrane prepared by a cementing together side by side hundreds of fine glass tubes, each of known cross section (DABROWSKI 1912), or of the porous sintered glass and alundum membranes used by NORTHROP and ANSON (1929) through which fairly large continuous aqueous channels are also available for diffusion. It would likewise be substantially true of so-called moist collodion membranes and of agar-agar or gelatin gels of not too great concentration[1]. It would not, however, in general, be

[1] It is impracticable in the present paper to discuss in detail the similarities and dissimilarities between diffusion processes in gels and in purely aqueous media. The similarities are frequently so great that the diffusion coefficients obtained in the two media differ only slightly from one another; in other cases, however, relatively great differences may appear. In general, the higher the molecular weight of the diffusing substances and the greater the concentration of the gel, the more will diffusion tend to be retarded as compared with its rate in water, but other factors are also involved and the details of the relation are by no means simple. The following references selected from the very voluminous literature on the subject will give access

even approximately true of dried collodion membranes or of tissues or of other media which interpose definite phase boundaries between the regions of known concentration and the region covered by the diffusion equation.

While it lies beyond the scope of the present paper to consider in more than a very superficial way processes of diffusion across phase boundaries, it may be useful merely to mention several of the various special problems that such boundaries raise. The first and simplest, as well as the most important practically, is that caused by a difference in the solubility of a diffusing substance in two dissimilar adjacent phases. This difference in solubility, except at a boundary at which a concentration of zero exists, gives rise to a differential distribution or partition between the two phases, which, over the relatively small ranges of concentration encountered in physiological phenomena, may usually be represented with sufficient accuracy by a constant "partition coefficient".

The importance of partition coefficients in discussions of cell permeability and related phenomena is too well known to require special emphasis at this point — see, for example, the recent papers by COLLANDER and BÄRLUND (1933) and OSTERHOUT (1933). It has sometimes been objected (BROOKS and BROOKS 1932) that since partition is an equilibrium phenomenon it cannot properly be used for the purpose of predicting rates of diffusion. In reply, however, it may be pointed out, as has been done by COLLANDER and BÄRLUND, OSTERHOUT and others, that the rate of diffusion across a non-aqueous phase depends on the difference rather than on the ratio of the concentrations of the diffusing substance on the two sides of this phase. If, therefore, the concentrations of the diffusing substance in the two aqueous media be $c_1$ and $c_2$, respectively, and the partition coefficient between the non-aqueous medium and water be $S$, the effective concentration gradient across the interposed layer, under equilibrium or nearly equilibrium conditions, must be $\frac{S c_1 - S c_2}{H}$ — in other words, the partition effect will multiply the gradient that would otherwise exist by $S$ and, other things being equal, will increase the rate of diffusion to the same

to much interesting information: GRAHAM (1861), CHABRY (1888), VOIGTLÄNDER (1889), REFORMATSKY (1891), PRINGSHEIM (1895a, b), HAGENBACH (1898), CALUGAREANU and HENRI (1901), NELL (1905), BECHHOLD and ZIEGLER (1906a, b), MEYER (1906), DUMANSKI (1908), YEGOUNOW (1906, 1908), ÖHOLM (1913), VANZETTI (1914), FÜRTH and BUBANOVIC (1918a, b), FÜRTH, BAUER and PIESCH (1919), GRAHAM and GRAHAM (1918), STILES (1920, 1921, 1923), STILES and ADAIR (1921), ADAIR (1920), TRAUBE and SHIKATA (1923), AUERBACH (1924), MANN (1924), FRICKE (1925), AFFONSKY (1928), FRIEDMAN (1930a, b), FRIEDMAN and KRAEMER (1930), RICKETTS and CULBERTSON (1931), EPPINGER and BRANDT (1932), HATSCHEK (1932), MAGISTRIS (1932), KLEMM and FRIEDMAN (1932).

extent. This point is further discussed by NORTHROP (1929) who used the principle in question in a discussion of the manner of passage of gases through dried collodion membranes. It is scarcely conceivable, even though partition equilibrium across phase boundaries were approached with only moderate rapidity, which seems to be a gross understatement of the case, that high values of $S$ could fail to have an important or even a predominant influence in determining rates of diffusion in systems characterized by such boundaries.

Because of the phenomenon of partition at phase boundaries it is obviously impossible to measure the true diffusion coefficient of a substance in, for example, a non-aqueous membrane separating two known aqueous media, or in an aqueous membrane separating two known gaseous phases, without a knowledge of partition or solubility coefficients for the phases in question. Sometimes these coefficients may be experimentally determined and then used in an appropiate manner to formulate the true boundary conditions for the region under investigation. At other times it is easier, or for other reasons preferable, to determine what are sometimes called permeability coefficients as contrasted with true diffusion coefficients.

Thus, in physiological work, it is more convenient to deal with the tensions of oxygen and other gases than with their actual concentrations in the liquids of the cells and tissues. For this reason the important numerical coefficients determined by KROGH (1919a) for various tissues are expressed in terms of the amount of gas in c.c. that in a given time would pass through an area of 1 $cm.^2$ with a concentration gradient represented by a difference in partial pressure of one atmosphere per micron. The true diffusion coefficient for the tissue in question could obviously be found at any time, by dividing KROGH's constant, after changing the unit of distance, by the solubility coefficient, $\alpha$, of the gas in the tissue, since a gradient of 1 atmosphere per cm. across the tissue is evidently the same as a gradient of $\alpha$ units of amount per $cm.^3$ per cm. of distance within the tissue itself. For example, HILL (1928) assuming that the solubility of oxygen in tissues is the same as that in water, i.e., 0.031 at $20^0$ C, obtained on dividing by this quantity KROGH's constant of $1.40 \times 10^{-5}$ a true diffusion coefficient of $4.5 \times 10^{-4}$ $cm.^2$/minute. In view of the fact that solubility coefficients were, for convenience, purposely omitted in the calculation of KROGH's constants, it is not surprising that these constants for different gases should bear to one another a numerical relation showing not only the expected influence of the square roots of the molecular weights of the different gases, but that of their solubilities in water as well. Exactly the same relation in essentially the same type of experiment had previously been found for different gases by EXNER (1875) and also appears in the physiological studies of TESCHENDORF (1924).

A second factor that must theoretically enter into the formulation of the proper mathematical boundary conditions for some given part of a heterogeneous system is the rate at which the molecules of the diffusing substance cross a given phase boundary and in this way tend to bring about equilibrium between adjacent layers on the two sides of the interface. This rate, in itself, seems to be enormously great, though it is always more or less obscured by the existence, in liquid or partially liquid systems, of unstirred layers within which the conditions are not those postulated for the body of liquid as a whole. For many years the question, of great interest to physiologists, has been discussed as to whether the "invasion" of a liquid by a gas, or the converse process of "evasion", is one which has a measurable rate distinct from that of the diffusion processes which always accompany it. BOHR (1899) obtained experimentally "evasion coefficients" indicating a relatively slow, or at all events a measurable, rate of escape of gases from water surfaces. KROGH (1910) later obtained a much more rapid rate which he believed might itself be too low.

Still later KROGH (1919a) made certain observations on the diffusion of gases through animal tissues which were opposed to the view that the time of invasion and evasion can be of appreciable magnitude. These observations were, first, that the same diffusion coefficient is obtained with membranes of different thicknesses and, second, that the same coefficient is obtained when the same membrane is exposed on the one hand to an atmosphere of gas, and on the other hand to a liquid previously brought into equilibrium with the same atmosphere. He also carried out special experiments which led him to the conclusion that it is impossible by existing methods to measure the rate of invasion, and that in any case the process is one which "is so rapid that it can be left out of account altogether in dealing with the rate of absorption of gases in animal tissues".

More recently, by a very beautiful and convincing method, DIRKEN and MOOK (1930) have shown that the time required for $CO_2$ to pass into the surface of a moving column of water is inappreciable, and that the time required for the absorption of a given amount is only that necessary for the diffusion of the gas into the interior of the liquid. GUYER and TOBLER (1934) have also come to the conclusion that the rate of escape of gases from relatively large liquid surfaces is determined almost solely by diffusion processes within the body of the liquid, and that evasion as such must occur almost instantly. On the other hand, McKAY (1930, 1932a) has found that while the absorption of moisture by leather immersed in water follows the simple diffusion laws, conditions are much more complex when the leather is exposed to a moist atmosphere. Even in cases involving the direct exposure of gels to water, McKAY (1932b) has reported significant departures from the classical laws of diffusion connected with special surface conditions, which he has investigated mathematically.

A third factor, of importance in heterogeneous systems containing liquids, namely, the presence of unstirred layers, has already been mentioned. The thickness of these layers, even with vigorous stirring, is surprisingly great. Thus, according to DAVIS and CRANDALL (1930), "for water stirred underneath at 1000 r.p.m. the effective thickness is about 0.0045 cm., which is over a hundred thousand times the diameter of the water molecule". According to the same investigators, solid particles of freshly precipitated $BaSO_4$ may be seen under the microscope, not only at the surface but for a short distance beneath it, to behave as if they were in "a separate medium of their own which slips along the water like a thin skin". (See in the same connection, LEWIS and WHITMAN 1924, and, in general, the extensive literature dealing with the problem of chemical reaction velocities in heterogeneous systems.)

To what extent unstirred layers will complicate a given diffusion problem depends of course on the nature of the system. If no phase boundaries are present, and if the diffusion coefficient is the same throughout the system, the effect of an unstirred layer is merely to increase the length of the region in which diffusion occurs by an amount which may or may not be significant, according to the relative proportions of the two regions. In heterogeneous systems, on the other hand, unstirred layers are of the greatest importance and may even determine the entire behavior of the system (see in this connection OSTERHOUT (1933). Their importance in connection with questions of invasion and evasion has already been mentioned.

It is evident, from this brief discussion that the complicating factors present in cases of diffusion in heterogeneous systems must greatly increase the difficulties encountered in dealing mathematically with physiological diffusion processes. Undoubtedly these factors will more and more be taken into account in future work, and for the present their existence and possible importance should never be forgotten. But since a large number of problems of physiological interest can be dealt with fairly successfully by methods applicable to simple homogeneous systems, and since these methods must in any case be thoroughly mastered before the use of others of greater complexity can profitably be undertaken, it is believed that the limitation of the present discussion to methods of the former type is not without justification.

## 6. Solution of the diffusion equation.

Before the diffusion equation can be put to practical use it must be solved; that is, a relation must be found between finite values of the variables which it contains. Like partial differential equations in general, equation (8) has many mathematical solutions, and the chief difficulty lies not so much in finding relations between the variables which satisfy it as in selecting from among all the possible relations of this sort the one which fits any particular diffusion problem. The simplest possible solution of the equation (8) arises when, in a homogeneous system in the absence of all external influences not provided for by FICK's law (such as for example gravitation), equilibrium is established. Under these conditions, $\frac{\partial u}{\partial x}$ is everywhere equal to zero, and we have as a first solution

$$u = c. \tag{11}$$

That (11) is in reality a solution of (8) is obvious from the fact that the substitution of $c$ for $u$ in the latter equation leads to an identity.

A second and almost equally simple solution arises in the case of a steady state where $\frac{\partial u}{\partial t} = 0$ but where $\frac{\partial u}{\partial x} \neq 0$. From the first of these two conditions we have (discontinuing the notation of partial differentiation when there are no longer two independent variables)

$$\frac{d^2 u}{d x^2} = 0.$$

This on integration gives

$$\frac{d u}{d x} = a$$

where $a$ is a constant which cannot have the value zero. On integrating a second time we have

$$u = a x + b. \tag{12}$$

That (12) is likewise a solution of (8) is again obvious from the fact that the substitution of $a x + b$ for $u$ in (8) leads to an identity, regardless of the values assigned to the constants $a$ and $b$. The particular values which these constants assume, however, in special problems is a matter of great importance. Suppose, for example, that the boundary conditions are $u = c_1$, when $x = 0$, and $u = c_2$ when $x = H$, $(c_1 > u > c_2)$; on making these substitutions in (12) we obtain the necessary values of the constants, and the equation becomes

$$u = c_1 - \frac{c_1 - c_2}{H} x. \tag{13}$$

This is the equation of a straight line; evidently the gradient represented by the coefficient of $x$ changes neither with time nor with distance and it may, therefore, be introduced into equation (1) even for finite times and quantities of diffusing material, thus

$$Q = \frac{D A (c_1 - c_2)}{H} t. \tag{14}$$

This equation is one of very great usefulness and will receive further attention below.

The two solutions of the diffusion equation so far obtained are extremely simple owing to the fact that in both cases the independent variable $t$ disappears, and the resulting ordinary differential equations in two variables can be solved by direct integration. When, as is usually the case, however, three variables must be dealt with simultaneously, the problem becomes more complex, and the special methods applicable to the solution of partial differential equations must be employed. The general method of procedure in such cases is to find certain particular mathematical solutions of the original equation which individually may not, and usually do not, fit the special problem under consideration but which, when properly combined, furnish the desired solution. Two steps are therefore involved: first, the finding of particular solutions and, second, the selection and proper combination of these solutions.

There is no fixed and infallible method for finding particular solutions of partial differential equations; in general, any method is permissible that leads to results which on trial are found to satisfy the original equation. In the present instance, we may in a purely tentative way assume that a solution of equation (8) may be found which has the form

$$u = XT \tag{15}$$

where $X$ is any function of $x$ alone and $T$ is any function of $t$ alone. If this hypothesis leads to a satisfactory solution, it may, like any other scientific hypothesis, be considered to be justified by its results; if not, it may be discarded in favor of another hypothesis.

On the assumption that (15) is a solution of (8) we may perform on the former the operations indicated in the latter, obtaining after a rearrangement of the terms

$$\frac{1}{DT}\frac{\partial T}{\partial t} = \frac{1}{X}\frac{\partial^2 X}{\partial x^2}. \tag{16}$$

Since, by hypothesis, the right-hand side of (16) does not involve $t$ nor the left-hand side $x$, the equation can be true for all values of $x$ and $t$ only if each side be equal to a constant, which cannot have the value zero without limiting the solution to the steady state which has already been considered. The constant in question may be represented by any desired symbol, and reasons will appear later for assigning to it the form $-\mu^2$. To save space this will be done at once and we have

$$\frac{dT}{dt} + \mu^2 DT = 0 \tag{17a}$$

and

$$\frac{d^2X}{dx^2} + \mu^2 X = 0. \tag{17b}$$

These differential equations, each involving only two variables, are of types whose solution is very easy by purely routine methods. By the application of these methods the following general solutions involving the proper number of arbitrary constants are obtained

$$T = Ce^{-\mu^2 Dt}$$

and

$$X = A \sin \mu x + B \cos \mu x.$$

For present purposes, however, it is best to break the second of these solutions into two particular solutions by placing $A$ and $B$ successively equal to zero. Substituting these latter values in (15) we have finally as hypothetical solutions of the diffusion equation

$$u = a \sin \mu x e^{-\mu^2 D t} \tag{18a}$$

$$u = b \cos \mu x e^{-\mu^2 D t}. \tag{18b}$$

A test of these two solutions shows that they do, in fact, satisfy equation (8), and further work with them will show that together with the solutions already found, they provide the necessary means for dealing with a great variety of problems.

It will be noted that (18a) and (18b) are equally valid as solutions of (8) for all values of $a$, $b$ and $\mu$. Furthermore, any number of these particular solutions added together are also a solution of (8) since the operations indicated in the latter equation, which is linear, do not in any way destroy the independence of the individual members of a series of such terms. We therefore have available an infinite number of sine-exponential and cosine-exponential terms differing from one another with respect to the constants $a$, $b$ and $\mu$. It remains to select from this infinite number of possibilities those appropriate to some given problem, and by combining them in a series to obtain the desired solution. The general method of procedure is first to select the types of terms that fit the two boundary conditions, and then by a further choice from among these selected terms to obtain a series which will satisfy the initial condition as well. Such a solution, in the form of an infinite series is, subject to the convergency of the series obtained, a complete solution of the problem.

This method of procedure may be illustrated by several examples. Consider first the boundary condition $u = 0$ when $x = 0$. It is obvious that since $\sin 0 = 0$ while $\cos 0 = 1$, terms involving sines rather than cosines will be needed in this case. On the other hand, for the boundary condition $\frac{\partial u}{\partial x} = 0$ when $x = 0$, it is equally obvious that cosines rather than sines are required. It was in order to permit this choice that the general solution of equation (17b) above, involving both sines and cosines, was broken up into two particular solutions in which the two functions appear separately.

Having satisfied the first boundary condition by selecting for further use either sine or cosine terms, as the case may be, the second boundary condition is next satisfied by an appropriate choice from among all the possible values of $\mu$. Suppose that $u = 0$, not only when $x = 0$, but when $x = H$ as well. By taking $\mu = \frac{n\pi}{H}$ where $n$ is any integer, $\sin \mu x$ must evidently become equal to zero when $x$ is given the value $H$. Both boundary conditions will therefore be satisfied simultaneously by a solution of the form

$$u = \sum_{n=1}^{n=\infty} \left( a_n \sin \frac{n\pi x}{H} e^{-\frac{n^2\pi^2 D t}{H^2}} \right) \tag{19}$$

where the summation symbol $\Sigma$ has its usual significance and where the constant coefficient $a_n$ of each term is yet to be determined. It will be noted that when $t = \infty$ this solution reduces to $u = 0$, as it should in view of the fact that equilibrium cannot be established until the internal concentration has reached that maintained at the two boundaries of the system. It was for this reason that in solving equations (17a) and (17b) above the constant was taken as having the form $-\mu^2$, which for all values of $\mu$ must have a negative sign and must therefore cause the resulting expression in equation (19) to vanish when $t = \infty$.

Suppose next that $\frac{\partial u}{\partial x} = 0$, not only for $x = 0$ but for $x = H$ as well. By exactly the same procedure as before a solution of the following type results

$$u = b_0 + \sum_{n=1}^{n=\infty} \left( b_n \cos \frac{n\pi x}{H} e^{-\frac{n^2\pi^2 D t}{H^2}} \right). \tag{20}$$

It will be observed that a constant term, $b_0$, has been introduced into this equation. The presence of such an added constant is of course mathematically permissible, since $u = b_0$ is itself a solution of the diffusion equation. In this case it is needed to show that when $t = \infty$, the equilibrium concentration is not zero, since by the nature of the two boundary conditions no material can escape from the system. The term, in fact, represents the average concentration of diffusible material, i.e., its total amount divided by the volume of the system. For an initial distribution $u = f(x)$

$$b_0 = \frac{1}{H} \int_0^H f(x)\, dx.$$

It may also be noted that whenever a general series of cosine terms is used such a constant term must be taken into account for mathematical reasons, since $b_n \cos \frac{n\pi x}{H}$ unlike $a_n \sin \frac{n\pi x}{H}$ does not reduce to zero when $n = 0$.

The case represented by the pair of boundary conditions $u = 0$ when $x = 0$; $\frac{\partial u}{\partial x} = 0$ when $x = H$ and that that represented by the pair, $\frac{\partial u}{\partial x} = 0$ when $x = 0$; $u = 0$ when $x = H$, are dealt with similarly, though a different choice of $\mu$ must be made. For example, suppose that terms containing $\sin \mu x$ have been selected in order that they may reduce to zero when $x = 0$. Since $\frac{\partial u}{\partial x}$ will involve $\cos \mu x$, it is necessary, if expressions of the latter type are to reduce to zero when $x = H$, to let $\mu = \frac{(2p+1)\pi}{2H}$ where $p$ can have the values 0, 1, 2, 3, etc. The solution of (8) in this case therefore assumes the form

$$u = \sum_{p=0}^{p=\infty} \left( a_n \sin \frac{(2p+1)\pi x}{2H} e^{-\frac{(2p+1)^2 \pi^2 D t}{4H^2}} \right). \tag{21}$$

Similarly, for the other case

$$u = \sum_{n=1}^{n=\infty} \left( b_n \cos \frac{(2p+1)\pi x}{2H} e^{-\frac{(2p+1)^2 \pi^2 D t}{4H^2}} \right). \tag{22}$$

In neither of these solutions does a constant term appear, since in both there is an open boundary at which the concentration zero is maintained, and the equilibrium concentration must therefore also be zero.

Having satisfied the first boundary condition by the selection of either sine- or cosine-containing terms, and the second boundary condition by choosing appropriate values of $\mu$, the initial condition still remains to be dealt with. To satisfy it we have at our disposal an infinite number of values, of $a_n$ and $b_n$ in equations (19) to (22) inclusive, from which any desired selection may be made. It will be noted that when $t = 0$ the exponential parts of all the terms disappear, and what is left is merely a series of sines or of cosines as the case may be, each term provided with a coefficient to which any desired constant value that meets our needs may be assigned. It is shown in detail in the special works on FOURIER's series (BYERLY 1893, EAGLE 1925, etc.) how the proper coefficients may be introduced into infinite series of sine or cosine terms like those in question to make such a series represent a given function over a given range. It will be sufficient here merely to indicate the general method of procedure.

In the first place, it may be noted that subject to certain conditions, discussed at length in the special treatises, and fulfilled in all cases likely to be of physiological interest, a given function can be represented between $x = 0$ and $x = H$ by either a series of sines of the form

$$\sum_{n=1}^{n=\infty} \left( a_n \sin \frac{n \pi x}{H} \right)$$

or of cosines of the form

$$\sum_{n=1}^{n=\infty} \left( b_n \cos \frac{n \pi x}{H} \right)$$

together in certain cases with an added constant. The form of the function between $-H$ and 0 will depend on whether sines or cosines are used, but since this region is of no significance in problems involving only positive values of $x$, its behavior need not in any way influence our choice as between sines and cosines. In general in the case of a sine series, the symmetry is such that within the limits $\pm H$, $f(x) = -f(-x)$, while for a cosine series within the same range $f(x) = f(-x)$. Since sine and cosine series are periodic in nature, with a period for the cases under discussion of $2H$, it of course follows that all conditions existing between $-H$ and $+H$ will be repeated between $+H$ and $+3H$ and so on.

Without attempting a formal proof that $f(x)$ in a given diffusion problem can be represented by a Fourier's series of sines or of cosines, respectively, let us assume this to be the case and proceed on such an assumption to find what values the coefficients of the successive terms must assume. As an illustration, the form assumed by equation (19) when $t = 0$ may be selected; in this case the coefficients $a_1$, $a_2$, etc., must be so chosen that

$$f(x) = a_1 \sin \frac{\pi x}{H} + a_2 \sin \frac{2\pi x}{H} + \ldots + a_m \sin \frac{m\pi x}{H} + a_n \sin \frac{n\pi x}{H} + \ldots$$

If we multiply each side of this equation by $\sin \frac{n\pi x}{H} dx$, where we may temporarily think of $n$ as being some particular integer, and integrate between 0 and $H$, all the integrals that appear fall under three general forms, namely,

$$\int_0^H f(x) \sin \frac{n\pi x}{H} dx \qquad a_m \int_0^H \sin \frac{m\pi x}{H} \sin \frac{n\pi x}{H} dx$$

and

$$a_n \int_0^H \sin^2 \frac{n\pi x}{H} dx.$$

Of these, the first can be evaluated only when $f(x)$ is known, so it will be allowed to stand without change. The second, as may readily be seen by consulting any table of standard integrals, must reduce to zero whatever the values of $m$ and $n$ may be, provided that they are dissimilar integers. The third integral, on the other hand, assumes the simple value $a_n H/2$. We therefore obtain as the appropriate value of $a_n$ for all integral values of $n$

$$a_n = \frac{2}{H} \int_0^H f(\lambda) \sin \frac{n\pi\lambda}{H} d\lambda$$

where, in order to avoid possible confusion later, a different variable of integration, $\lambda$, which does not in any way alter the value of the definite integral, has been substituted for $x$. On introducing this value of $a_n$ in equation (19) we obtain a complete solution of the problem under consideration, namely,

$$u = \frac{2}{H} \sum_{n=1}^{n=\infty} \left( e^{-\frac{n^2\pi^2 Dt}{H^2}} \sin \frac{n\pi x}{H} \int_0^H f(\lambda) \sin \frac{n\pi\lambda}{H} d\lambda \right). \qquad (23)$$

This equation, as it should, satisfies the fundamental diffusion equation and the two boundary conditions for all values of $t$ and all values of $x$ lying between 0 and $H$, while for $t = 0$ it reduces to a FOURIER's series equal to $f(x)$, thereby satisfying the initial equation as well. It is therefore a complete solution of the problem.

The second case, involving a cosine series, is dealt with in exactly the same way except that the FOURIER's cosine series contains a constant term, $b_0$, which has already been introduced into the equation for purely physical reasons. In order to represent $f(x)$ by a cosine series all values of $b_n$ are obtained by multiplying both sides of the equation

$$f(x) = b_0 + b_1 \cos\frac{\pi x}{H} + b_2 \cos\frac{2\pi x}{H} + \ldots + b_m \cos\frac{m\pi x}{H} + b_n \cos\frac{n\pi x}{H} + \ldots$$

by $\cos\frac{n\pi x}{H}\,dx$ and integrating as before between 0 and $H$. The integrals now obtained fall into the four types

$$\int_0^H f(x)\cos\frac{n\pi x}{H}\,dx, \qquad b_0\int_0^H \cos\frac{n\pi x}{H}\,dx, \qquad b_m\int_0^H \cos\frac{m\pi x}{H}\cos\frac{n\pi x}{H}\,dx,$$

and

$$b_n\int_0^H \cos^2\frac{n\pi x}{H}\,dx.$$

The second and third of these types reduce to zero and we have, as before

$$b_n = \frac{2}{H}\int_0^H f(\lambda)\cos\frac{n\pi\lambda}{H}\,d\lambda.$$

To find $b_0$, both sides of the equation are multiplied by $dx$ and integrated between 0 and $H$. All integrals of the type

$$b_n\int_0^H \cos\frac{n\pi x}{H}\,dx$$

reduce to zero, while that associated with $b_0$ takes the value $H$, and we therefore have

$$b_0 = \frac{1}{H}\int_0^H f(\lambda)\,d\lambda.$$

This value is evidently the same as that given above for the average concentration existing within the system when $t = 0$; it is also the equilibrium concentration when $t = \infty$. Introducing the appropriate values of $b_0$ and $b_n$ into equation (20) above, we obtain finally, as a complete solution of the second case, which satisfies all the conditions of the problem

$$u = \frac{1}{H}\int_0^H f(\lambda)\,d\lambda + \frac{2}{H}\sum_{n=1}^{n=\infty}\left(e^{-\frac{n^2\pi^2 Dt}{H^2}}\cos\frac{n\pi x}{H}\int_0^H f(\lambda)\cos\frac{n\pi\lambda}{H}\,d\lambda\right). \quad (24)$$

The two cases mentioned above, each involving one open and one closed boundary may, if desired, be reduced to a single one merely by reversing the direction in which $x$ is measured, i. e., by exchanging the positions of $x = 0$ and $x = H$. For completeness, however, it will be useful to treat the two cases separately, beginning with the one having the boundary conditions $u = 0$ when $x = 0$ and $\frac{\partial u}{\partial x} = 0$ when $x = H$. For this case, the solution, as has already been shown, has the form of equation (21) above. It will be observed that the sine series that results when $t = 0$, unlike those already considered, involves terms of the form $\sin \frac{m \pi x}{2H}$ in which $m$ can have only odd values. To find the coefficients of a FOURIER's series of this type requires a procedure slightly different from already described. To change the denominator, $H$, to $2H$, it is merely necessary to make the length of the period twice as great as before, that is, to find a FOURIER's series for $f(x)$ between $-2H$ and $+2H$ instead of between $-H$ and $+H$ The elimination of the even terms of the series must next be accomplished. It has already been mentioned that in problems such as those under discussion, the only physically significant part of the period represented by the FOURIER's series is that from 0 to $H$; it follows, therefore, that not only may the region from $-H$ to 0 be treated in a way that corresponds to no physical reality, as has already been done by arbitrarily choosing a series consisting of either sines or cosines alone, but that the region from $H$ to $2H$ is also at our disposal in the same way. Advantage may be taken of this fact to eliminate the even terms of the complete series. Let the artificial mathematical function over the entire range $-2H$ to $2H$ be represented by $F(x)$. Between 0 and $H$ it is to coincide with $f(x)$ but need not do so elsewhere. By exactly the same methods as those previously used, with the substitution of $F(x)$ for $f(x)$, $2H$ for $H$ and $m$ for $n$ we obtain

$$a_n = \frac{1}{H} \int_0^{2H} F(\lambda) \sin \frac{m \pi \lambda}{2H} \, d\lambda .$$

This may be written

$$a_n = \frac{1}{H} \int_0^{H} f(\lambda) \sin \frac{m \pi \lambda}{2H} \, d\lambda + \frac{1}{H} \int_H^{2H} g(\lambda) \sin \frac{m \pi \lambda}{2H} \, d\lambda . \tag{25}$$

Consider now the nature of $\sin \frac{m \pi x}{2H}$. By analogy with $\sin \frac{n \pi x}{H}$ which has $n$ complete periods between $-H$ and $+H$ it must have $m$ complete periods between $-2H$ and $+2H$. If $m$ be odd, then the curve represented by the sine function can easily be seen by constructing a simple graph to have an even symmetry about $\lambda = H$, that is, to have the same values for $\lambda$ and for $2H - \lambda$. If, on the other hand, $m$ be even, then the symmetry about $H$ is of such a nature that the values corresponding to $\lambda$ and to $2H - \lambda$ are equal but of opposite sign. Now choose $g(\lambda)$ between $H$ and $2H$ so as to make it equal to $f(2H-\lambda)$. It is evident that under these conditions the second integral in (25) will be exactly equal to the first when $m$ is odd, but that it will be equal and of opposite sign, and the two will therefore cancel, when $m$ is even. By this arbitrary choice of a function, therefore, within a region that has no physical significance, a FOURIER's series having the desired properties over the significant region may be obtained; and representing odd values of $m$, as before by $2p+1$, the complete solution for this case becomes

$$u=\frac{2}{H}\sum_{p=0}^{p=\infty}\left(e^{-\frac{(2p+1)^2\pi^2Dt}{4H^2}}\sin\frac{(2p+1)\pi x}{2H}\int_0^H f(\lambda)\sin\frac{(2p+1)\pi\lambda}{2H}\,d\lambda\right). \quad (26)$$

Finally, for completeness the equation may be given for the case where $\frac{\partial u}{\partial x}=0$ when $x=0$, and $u=0$ when $x=H$. Its derivation is exactly the same as that of (26) except that cosine terms must now be used, while to obtain a series involving only odd values of $m$, $g(\lambda)$ must be taken equal to $-f(2H-\lambda)$. The equation so derived is

$$u=\frac{2}{H}\sum_{p=0}^{p=\infty}\left(e^{-\frac{(2p+1)^2\pi^2Dt}{4H^2}}\cos\frac{(2p+1)\pi x}{2H}\int_0^H f(\lambda)\cos\frac{(2p+1)\pi\lambda}{2H}\,d\lambda\right). \quad (27)$$

Equations (23) (24) (26) and (27), which represent general solutions of the diffusion equation for any desired initial condition and for four important pairs of boundary conditions, will now be applied in greater detail to various concrete problems of practical importance.

## 7. One-dimensional diffusion processes in closed systems.

Among the various methods available for the measurement of diffusion coefficients, the ones most commonly employed in the past have been those in which diffusion takes place within a closed system of fairly limited extent. These methods, which differ considerably in their details, nearly all start with the same type of initial distribution, namely, a layer of solution in a vessel of uniform cross section overlaid by another layer of pure solvent. The most frequent proportions for the depths of the layers of solution and solvent, respectively, are 1 : 1 and 1 : 4, though in the early experiments of GRAHAM (1861), who was the first to use the method, they were 1 : 8; and other proportions are occasionally mentioned in the literature. The subsequent behavior of such systems may be followed in a large number of different ways, all of which must be supplemented by appropriate mathematical treatment before the observed results can be put to any very practical use. It will be advantageous, therefore, to derive the equations necessary for dealing with such cases, after first considering briefly certain further details concerning the more important experimental methods.

In general, observations on diffusion in closed systems take the form either of measurements of the concentrations existing at given times and at given levels, or of the amounts of substance contained at some given time in regions lying between two chosen levels. The former type of measurement can be repeated at will throughout the period of observation; the latter type is usually made once for all at the conclusion of an experiment. Corresponding to these two types of measurement, two kinds of equations will be needed, the first relating the concentration $u$ to the diffusion coefficient $D$ and the variables $x$ and $t$, and the second relating the amount of substance, $Q_{x_1, x_2}$, lying between the levels $x_1$ and $x_2$, to $D$ and $t$. Rarely, in the case of closed systems, an equation may be useful which permits the calculation of the amount, $Q_{0, t}$, which would diffuse across some chosen level in the system within a given time.

Of the numerous methods which permit the observation of changes in concentration in situ without the interruption of the experiment, the most important are optical in character. When the diffusing substance is colored or may be made visible by means of fluorescence or by the absorption of ultraviolet light (SVEDBERG 1925), the general procedure is obvious. For other substances, an early suggestion made by SIMMLER and WILD (1857) and actually carried out by VOIT (1867) and JOHANNISJANZ (1877) was to allow diffusion to occur in a prism-shaped vessel and to estimate the concentrations at different levels by the refraction of a horizontal beam of light. Unfortunately, this method failed to take into account the fact, noted many years earlier by WOLLASTON (1800) and later emphasized by STEFAN (1878b), that in a system showing concentration differences there is a bending of a transverse beam of light in the direction of the region of greater concentration. This bending at a given level is approximately proportional to the concentration gradient at that level. While this circumstance prevents the use of the simple refraction method first suggested, it makes possible another method which has been rather extensively used (WIENER 1893, THOVERT 1901—1914, HEIMBRODT 1904). More recently, refraction methods have also been employed by LITTLEWOOD (1922), CLACK (1924), LAMM (1928) and MÜNTER (1931). Attempts to measure concentrations at different levels by the methods of polarimetry were made by HOPPE-SEYLER (1866) and by VOIT (1867) but are open to the same objections as the earliest applications of the refraction method. Perhaps the most satisfactory of all optical methods is the one recently introduced by ZUBER (1932) in which advantage is taken of the relation between the concentration of a solution and the occurrence of total reflection of light in an appropriately constructed diffusion vessel.

Another group of methods for following concentration changes in situ employ electrical measurements of various sorts. Thus, WEBER

(1879) used the changes in the potential difference between two metal electrodes in a solution of a salt of this metal as a quantitative measure of diffusion. His method has, however, been criticized by SEITZ (1898) and by HOELTZENBEIN (1924). PROCOPIU (1918) also employed a different form of the potential difference method. Another type of electrical method depends upon measurements of electrical conductance between appropriately placed electrodes (HASKELL 1908, MINES 1910).

A third group of methods depends on the changes in density that occur within the liquid medium during the course of diffusion. An early but crude application of this principle was made by FICK (1855), who used a glass bead suspended from the beam of a balance to measure specific gravities at different levels in his diffusion systems. A better method, since it avoids convection currents, is to introduce into the system at the beginning of the experiment a number of small floats of different specific gravities; the subsequent positions of these floats give a fairly accurate picture of the course of diffusion. Among those who have employed this method are THOULET (1891), WILKE and STRATHMEYER (1926) and GERLACH (1931). This list by no means exhausts the methods available for studying diffusion in situ, but is at least sufficient to illustrate their variety. For fuller details and for additional references to the literature FÜRTH (1931a) and WILLIAMS and CADY (1934) may be consulted.

Though, in theory, methods involving a considerable number of observations of concentration are preferable to those in which a single chemical analysis, or other measurement of quantity terminates a given experiment, methods of the latter type have been much more used in the past, chiefly because of their greater simplicity. Beginning with GRAHAM (1861) and continuing almost to the present day, by far the most commonly employed method for the quantitative study of diffusion has been the so-called method of layers. In this method, after diffusion has progressed for some suitable time, the entire body of liquid is separated into a convenient number of layers, which are then subjected to chemical analysis, and the general distribution of the material in the system is thus determined.

In GRAHAM's original experiments, the observed data were not given any further mathematical treatment, and a fairly large number of layers (16) was thought necessary to give an accurate picture of the character of the diffusion process. Following the theoretical discussion of these results by STEFAN (1879), however, it was realized that fewer layers suffice for the determination of a diffusion coefficient, which concisely and quantitatively describes the nature of the process. A very early method involving only two layers was that used by LOSCHMIDT (1870a, b), STEFAN (1871), and OBERMAYER (1880—1887) in the study of the diffusion of gases, but while it is still sometimes employed in studies of diffusion in solutions, it has been largely replaced by a standardized four-layer method which is somewhat more accurate. A further reason for the greater popularity of the four-layer method, which has been much more frequently used than any other diffusion method, was the preparation by SCHEFFER (1888) and particularly by KAWALKI (1894) from calculations originally made by STEFAN (1879) of tables to fit this particular case. By means of these tables, diffusion coefficients can be determined directly from the observations with a minimum of mathematical labor. Among the large number of workers who have employed the four-layer method and who, with the exception of the first three mentioned, have evaluated their results by means of KAWALKI's tables may be mentioned: SCHEFFER (1888), ARRHENIUS (1892), ABEGG (1893), KAWALKI (1894, 1896), ÖHOLM (1905—1913), HERZOG (1907a, b), HERZOG and KASARNOUSKI (1908), SVEDBERG (1909), SVEDBERG and ANDREEN-SVEDBERG (1911), v. HEVESY (1913a, b), HERZOG and POLOTZKY (1914), RONA (1918, 1920), COHEN and BRUINS (1923a, b, 1924), MILLER (1924), JANDER and SCHULTZ (1925), GROH and KELP (1925), MUCHIN and FAERMANN (1926), BURRAGE (1932), etc. The papers of these workers may be consulted for practical details as to the best methods of bringing together and separating the various layers of liquids, etc.

In dealing with observations of the continuous type, an equation is needed which gives for any value of $D$ the relation between finite values of $u$, $x$, and $t$; or which, conversely, enables $D$ to be calculated from observed values of these three variables. Such an equation, for a system in which a layer of solution of depth $h$ is initially covered by a layer of water of depth $H - h$, may readily be obtained from the more general equation for closed systems [(24), p. 37]. It is only necessary to introduce into the latter equation the proper initial condition $u = u_0$ from $x = 0$ to $x = h$ and $u = 0$ from $x = h$ to $x = H$ when $t = 0$. (If desired, $x$ may be measured from the top instead of the bottom of the vessel; in that case the problem is treated in essentially the same manner, though slightly different equations will be obtained.)

Since the initial distribution of diffusible material within the system is discontinuous, two separate integrals must be used to cover the range from 0 to $H$; and as the second one, with limits $h$ and $H$, has the value zero, we obtain immediately

$$u = \frac{u_0}{H}\int_0^h d\lambda + \frac{2u_0}{H}\sum_{n=1}^{n=\infty}\left(e^{-\frac{n^2\pi^2 Dt}{H^2}}\cos\frac{n\pi x}{H}\int_0^h \cos\frac{n\pi\lambda}{H}\,d\lambda\right)$$

which after the integrations have been performed becomes

$$u = \frac{u_0 h}{H} + \frac{2u_0}{\pi}\sum_{n=1}^{n=\infty}\left(\frac{1}{n}\sin\frac{n\pi h}{H}\cos\frac{n\pi x}{H}e^{-\frac{n^2\pi^2 Dt}{H^2}}\right). \qquad (28)$$

From this equation $u$ may be calculated for any given values of $x$, $t$ and $D$. It will be noted that when $t = \infty$, $u$ assumes, as it should, the average equilibrium value obtained by dividing the amount of solute, initially present in the layer of solution, i. e., $u_0 h A$, by the volume of the entire system, $HA$, where $A$ is the cross section of the vessel.

The usefulness of equation (28) will of course depend upon the rapidity with which the infinite series of trigonometric-exponential terms converges; obviously large values of $t$ and $D$ and small ones of $H$ will favor convergence. In other words, the more nearly the diffusion process has been completed, whether because the solute is one that diffuses rapidly, or because the time that has elapsed has been long, or because the distance to be covered is small, or because of any combination of these three factors, the easier it will be to employ equation (28). Fortunately, in cases where the effect of these factors together would give a slowly converging series, i. e., when the process has proceeded only a small part of the way towards its completion, it is permissible to treat the system as if the layer of solvent were infinite in height. The necessary calculations may then be made with ease by a method to be discussed below (p. 92). It is always possible, therefore, by one method of calculation or the other, to obtain with any desired degree of accuracy the value of $u$ corresponding to any given values of $x$, $t$ and $D$.

It should be noted that by the proper choice of experimental conditions it is frequently possible to save considerable mathematical labor. Thus, if $h$ in equation (28) be taken as equal to $H/2$, i. e., if columns of equal thickness of solution and water be employed, then $\sin \frac{n \pi h}{H}$ assumes the simple series of values: 1, 0, — 1, 0, 1, etc., and half of the terms of the resulting equation disappear, with a correspondingly more rapid convergence of the series. The equation for this particular, very important case may be written in the form

$$u = \frac{u_0}{2} + \frac{2 u_0}{\pi} \left( \cos \frac{\pi x}{H} e^{-\frac{\pi^2 D t}{H^2}} - \frac{1}{3} \cos \frac{3 \pi x}{H} e^{-\frac{9 \pi^2 D t}{H^2}} + \ldots \right). \tag{29}$$

It may also be noted that starting with equal columns of solution and of water, if observations of $u$ be made, not at random but at certain properly selected levels, an even greater mathematical simplification of equation (28) results. Thus for $x = H/6$ and $x = 5\,H/6$, the third as well as the second term of the original series disappears, because of the fact that $\cos \frac{\pi}{2}$ and $\cos \frac{5\pi}{2}$ are each equal to zero. Since the fourth term of the original series disappears for the same reason as the second, and since the fifth and subsequent terms are very small, except during the earliest stages of the process, the first term alone may, under these conditions, frequently be sufficient for all practical purposes of calculation.

It will also be apparent from an examination of equation (29) that if observations be made at the level $x = H/2$, i. e., at the original junction of the two solutions, a concentration of $u_0/2$ ought to be found at that level for all values of $t$, since under these conditions all the cosine terms assume the value zero. The same result is obtained in the consideration of infinite systems (p. 95), and has certain important practical consequences which will be mentioned later (p. 131). With finite systems, this relation will of course not be obtained when the initial layers of solution and water are of unequal thickness.

While equations (28) and (29) enable all cases to be dealt with which involve measurements of concentrations at different levels and times, it has already been mentioned that the commonest closed-system methods depend upon chemical determinations at the end of the experiment of the quantities of the diffusing substance contained in layers of finite thickness. The equations necessary to evaluate this type of experimental data are very readily derived. It is obvious that the amount of substance contained in an elementary layer lying between $x$ and $x + dx$ is $u A dx$, where $A$ is the area of the layer, and where $u$ can be obtained from equation (28). The total amount lying between any two finite values of $x$ can therefore be found by the integration of $u A dx$ between the values of $x$ in question. In order to have an equation that will be generally applicable, it is best to use 0 and $x$ as the limits of integration; the amount, $Q_{x_1, x_2}$, lying between the levels $x_1$ and $x_2$ can then always be obtained by subtracting $Q_{0, x_1}$ from $Q_{0, x_2}$.

The necessary integration presents no difficulties, since each term of the series in equation (28) is independent of all the others and may be integrated separately after having been multiplied by $A dx$; $t$, of course, under the conditions in question behaves as a constant. The equation obtained in this manner is

$$Q_{0,x} = \frac{u_0 A h x}{H} + \frac{2 u_0 A H}{\pi^2} \sum_{n=1}^{n=\infty} \left( \frac{1}{n^2} \sin \frac{n \pi h}{H} \sin \frac{n \pi x}{H} e^{-\frac{n^2 \pi^2 D t}{H^2}} \right). \qquad (30)$$

From this equation might be calculated such a table as that of KAWALKI (1894), though in actual practice this particular calculation may more easily be made, and was in fact made by STEFAN (1879), by a different method (see p. 93).

Equation (30) assumes a considerably simpler form when layers of equal thickness of solution and of solvent are employed, and when the final separation for chemical analysis is also made along the original boundary plane between these two layers. For this particular case, in which $h = H/2$, and $x = H/2$ equation (30) becomes

$$Q = \frac{u_0 A H}{4} \left[ 1 \pm \frac{8}{\pi^2} \left( e^{-\frac{\pi^2 D t}{H^2}} + \frac{1}{9} e^{-\frac{9 \pi^2 D t}{H^2}} + \dots \right) \right], \qquad (31)$$

where the positive sign applies to the lower, and the negative sign to the upper half of the vessel. It will be seen from this equation that when $t = \infty$, i.e., at equilibrium, the amounts of diffusible material in the two halves of the system, as they should be, are the same, and are equal in each case to half of the total amount of substance originally present in the system. Furthermore, when $t = 0$ $\left(\text{since } 1 + \frac{1}{9} + \frac{1}{25} + \ldots = \frac{\pi^2}{8}\right)$, the amount in the lower half of the vessel is equal to $\frac{u_0 A H}{2}$ and that in the upper half to zero—again as they should be.

A yet simpler equation may be obtained by dividing the difference between the amounts of substance in the two halves, of the system, i. e., $Q_1 - Q_2$, by their sum, $Q_1 + Q_2$, the latter being the total amount of substance present in the system. The value of this ratio gives rise to the useful equation

$$\frac{Q_1 - Q_2}{Q_1 + Q_2} = \frac{8}{\pi^2}\left(e^{-\frac{\pi^2 D t}{H^2}} + \frac{1}{9} e^{-\frac{9\pi^2 D t}{H^2}} + \ldots\right). \tag{32}$$

For all except very small values of $Dt/H^2$ the series of exponential terms on the right-hand side of the equation converges so rapidly that only the first member need be taken into account. Under these conditions the following very simple equation becomes available for the calculation of $D$ from experimental data

$$D = \frac{H^2}{\pi^2 t} \ln \frac{8(Q_1 + Q_2)}{\pi^2 (Q_1 - Q_2)}. \tag{33}$$

Equation (32) was first employed by LOSCHMIDT (1870a, b) in studies on the diffusion of gases and was later used, among others, by STEFAN (1871), OBERMAYER (1880—1887), RAMSTEDT (1919) and apparently by EDGAR and DIGGS (1916), though an error seems to have been made in the equation actually published by the latter authors.

| $\frac{Dt}{H^2}$ | $\frac{8}{\pi^2}\left(e^{-\frac{\pi^2 Dt}{H^2}} + \frac{1}{9} e^{-\frac{9\pi^2 Dt}{H^2}} + \ldots\right)$ | $\frac{Dt}{H^2}$ | $\frac{8}{\pi^2}\left(e^{-\frac{\pi^2 Dt}{H^2}} + \frac{1}{9} e^{-\frac{9\pi^2 Dt}{H^2}} + \ldots\right)$ |
|---|---|---|---|
| 0.001 | 0.929 | 0.04 | 0.549 |
| 0.002 | 0.899 | 0.05 | 0.486 |
| 0.003 | 0.876 | 0.06 | 0.449 |
| 0.004 | 0.857 | 0.07 | 0.406 |
| 0.005 | 0.840 | 0.08 | 0.368 |
| 0.006 | 0.825 | 0.09 | 0.333 |
| 0.007 | 0.811 | 0.1 | 0.302 |
| 0.008 | 0.798 | 0.2 | 0.113 |
| 0.009 | 0.785 | 0.3 | 0.042 |
| 0.01 | 0.774 | 0.4 | 0.016 |
| 0.02 | 0.681 | 0.5 | 0.006 |
| 0.03 | 0.609 | | |

When, for any reason, it is necessary to use more than the first exponential term appearing in equations (31) and (32), a table of values of $1-\frac{8}{\pi^2}\left(e^{-v}+\frac{1}{9}e^{-9v}+\ldots\right)$ prepared by McKAY (1930) for values of $v$ ranging by steps of 0.001 from $v=0$ to $v=3.509$ will be found to save much unnecessary labor. In the absence of a more elaborate table the very brief one given above will sometimes be helpful for making rough calculations.

Reference was made above to the optical methods which involve the determination of values of $\frac{\partial u}{\partial x}$ by measuring the bending of a transverse beam of light towards the more concentrated part of the solution. The use of such methods requires a knowledge of the relation between $\frac{\partial u}{\partial x}$ and $x$, $t$ and $D$. This is readily found by differentiating equation (28) with respect to $x$, giving

$$\frac{\partial u}{\partial x}=-\frac{2u_0}{H}\sum_{n=1}^{n=\infty}\left(\sin\frac{n\pi h}{H}\sin\frac{n\pi x}{H}e^{-\frac{n^2\pi^2 Dt}{H^2}}\right). \tag{34}$$

As was pointed out by THOVERT (1901), this equation may be made more useful by taking $h=H/2$ and by measuring $\frac{\partial u}{\partial x}$ either at $x=H/3$ or $x=2H/3$. In both cases, the second third and fourth terms of the original series drop out, and the fifth and succeeding terms are small enough to be neglected, except for very small values of $t$. A simple equation therefore results from which $D$ may readily be calculated. By a mathematical artifice the series can be reduced, practically speaking, to a single term, even for very small values of $t$. The artifice in question is to take the sum of the values of $\frac{\partial u}{\partial x}$ for $x=H/2$ and $x=H/6$, respectively, and to add to this sum $\sqrt{3}$ times the corresponding value for $x=H/3$. From the total, not only the terms mentioned, but all terms through the tenth of the original series also disappear, leaving after the first term no other until $\frac{1}{121}e^{-\frac{121\pi^2 Dt}{H^2}}$ is reached; this and subsequent terms are almost always negligibly small.

Having the value of $\frac{\partial u}{\partial x}$, it is very easy to calculate the total amount of substance that in a given time would cross any plane at right angles to the direction of diffusion. By FICK's law, the amount, $dQ$, that would pass the level $x$ in time $dt$ is obtained by multiplying equation (34) by $-DA\,dt$. The amount $Q_{0,t}$ that would pass between the beginning of an experiment and the time $t$ is then obtained by integrating the resulting expression term by term between 0 and $t$, $x$ in this case being treated as a constant. The equation so obtained is

$$Q_{0,t}=\frac{2u_0AH}{\pi^2}\sum_{n=1}^{n=\infty}\left[\frac{1}{n^2}\sin\frac{n\pi h}{H}\sin\frac{n\pi x}{H}\left(1-e^{-\frac{n^2\pi^2 Dt}{H^2}}\right)\right] \tag{35}$$

This equation could equally well be used to obtain the same results as those given, for example, by equation (31) above. It would merely be necessary in this case to subtract from or to add to the amount of solute originally contained in either half of the vessel the amount shown by equation (35) to have crossed the boundary plane between the two halves in time $t$.

## 8. One-dimensional diffusion processes in systems with one open boundary.

Though closed systems of the type described in the preceding section are of much practical importance in the determination of diffusion coefficients, they are of less interest to the physiologist than those in which one or both boundaries are "open", i. e., in which they are capable of being actually crossed by diffusing materials. Though, logically, systems with one and with two open boundaries seem sufficiently distinct, they are in reality very closely related, and many of the equations derived for one type of system may, with slight changes, be applied to the other as well. For this reason, both types are frequently treated together, as will be done here to some extent, though for mathematical and other reasons a formal separation of the two will be preserved.

It is of interest to note that the earliest method employed for the quantitative study of free diffusion, namely, the so-called first method of Graham (1850 a), involved the use of a system with one open and one closed boundary. In the simplest form of this method, a small vessel with an open top is filled with a solution of the substance whose diffusion is to be measured and the vessel is then surrounded and covered by a large quantity of water contained in an outer vessel. Because of the higher specific gravity of the solution, no mixing theoretically occurs except by diffusion, and, furthermore, since materials of greater density must settle to the bottom of large vessel, the top of the small vessel is kept in contact with practically pure water at all times. The boundary condition for this surface of the system is therefore $u = 0$ when $x = H$, or $u = 0$ when $x = 0$, according to whether distance is measured from the bottom or from the top of the vessel. The other boundary condition is evidently $\frac{\partial u}{\partial x} = 0$, since no material can cross the bottom of the vessel. The initial condition may represent any distribution of the diffusible substance in the inner vessel at the beginning of the experiment. Frequently the initial concentration is uniform throughout the vessel, i. e., $f(x) = u_0$ when $t = 0$, though Graham found it practically more convenient to fill the inner vessel partly with solution and partly with water; and Simmler and Wild (1857) pointed out certain mathematical advantages of this arrangement, provided that the quantities of solution and solvent be properly chosen.

In GRAHAM's experiments, and in the somewhat similar ones of MARIGNAC (1874), PICKERING (1893) and several other workers, the rates of diffusion of different substances were estimated merely by determining what proportions of their original amounts remained in the inner vessel at the end of some arbitrarily chosen time. Obviously, the more rapid the rate of diffusion, the greater must be the loss in a given time; and by using a series of similar vessels containing different substances and determining by chemical analysis the amount of each that after a given time still remains in its respective vessel, the substances can be arranged in the order of their diffusion rates. It is incorrect, however, to assume, as has sometimes been done, that the amounts of the different substances escaping in this manner furnish a quantitative measure of fundamental diffusion rates. This point will be dealt with in a more mathematical manner below, but the fallacy of such an assumption is obvious from the fact that if very long times be selected for the comparison—and the choice of the time for a given experiment is purely arbitrary—all substances must show approximately the same loss from the inner vessel, namely, nearly the total amount present in each case. The longer the time chosen for comparison, therefore, the more nearly must all substances, if compared by this invalid method, appear to diffuse at the same rate.

Data so obtained, however, when mathematically treated in the proper manner, are capable of yielding fairly accurate diffusion coefficients. Such a treatment was first furnished by SIMMLER and WILD (1857) soon after GRAHAM had made the experiments mentioned above, though unfortunately it was not applicable to GRAHAM's experiments because his small diffusion vessels were not uniform in cross section. The equations derived by SIMMLER and WILD have, however, been used to very good advantage by SCHEFFER (1882, 1883), SCHEFFER and SCHEFFER (1916) and others, who have employed GRAHAM's first method in a somewhat improved form. The same equations have also frequently been applied to the case of diffusion in either direction between liquids of constant composition and masses of agar-agar or other water-saturated gels contained in tubes closed at one end. [A more complicated, but practically important, case in which the composition of the liquid in contact with the gel is not constant but is gradually increased by outward diffusion has been treated by MARCH and WEAVER (1928).]

The general diffusion equations for systems with one open and one closed surface have been given above [numbers (27) and (26)]. It remains merely to introduce into these equations the appropriate initial conditions. For the most important case in which $f(x) = u_0$ when $t = 0$, the equations in question lead to

$$u = \frac{4\,u_0}{\pi} \sum_{p=0}^{p=\infty} \left( \frac{(-1)^p}{2\,p+1} \cos \frac{(2\,p+1)\,\pi\,x}{2\,H} \, e^{-\frac{(2\,p+1)^2 \pi^2 D t}{4\,H^2}} \right) \tag{36}$$

and

$$u = \frac{4\,u_0}{\pi} \sum_{p=0}^{p=\infty} \left( \frac{1}{2\,p+1} \sin \frac{(2\,p+1)\,\pi\,x}{2\,H} \, e^{-\frac{(2\,p+1)^2 \pi^2 D t}{4\,H^2}} \right) \tag{37}$$

respectively, according to whether the bottom or the top of the vessel is chosen as the origin for the measurement of $x$, i. e., according to whether the closed boundary is represented by $x = 0$ or by $x = H$. From either equation the concentration at any level of the inner vessel may be found for any desired values of $x$, $t$ and $D$. Because of the general similarity of these two cases, further mathematical treatment will here be based upon equation (36) only, though corresponding methods could equally readily be applied to the second one.

Before proceeding farther it is important to note that equations (27) and (36), though derived for cases involving only one open boundary, may equally well be used for systems with two such boundaries, provided that the two boundary conditions are identical and that the initial distribution of diffusible material is symmetrical about the mid-plane of the system. This follows from the fact that these equations involve only cosine terms, and $\cos x = \cos(-x)$. If, therefore, the origin be taken at the mid-plane of a system of thickness $2\,H$, equation (36), enables $u$ to be calculated for all values of $x$ lying between $-H$ and $+H$. This is, in fact, the form of the equation for a system with two open boundaries preferred by many investigators, and several examples of its use in connection with physiological problems will be given below.

In systems of this sort, it is frequently advantageous to be able to find the rate at which the solute escapes from the inner vessel at any time $t$. This rate is evidently equal to the concentration gradient at the open boundary multiplied by $-DA$. The gradient in question is found by differentiating either (36) or (37) term by term with respect to $x$ and substituting $x = H$ or $x = 0$, as the case may be. The result, except for sign, is the same in both cases, namely,

$$\frac{\partial u}{\partial x} = \mp \frac{2\,u_0}{H} \left( e^{-\frac{\pi^2 D t}{4\,H^2}} + e^{-\frac{9\,\pi^2 D t}{4\,H^2}} + \cdots \right). \tag{38}$$

The total amount that would escape between the times $t = 0$ and $t = t$ is then obtained by integrating equation (38) between 0 and $t$, after first having multiplied it by $-DA\,dt$ or by $DA\,dt$, as the case may be, remembering that $1 + \frac{1}{9} + \frac{1}{25} + \cdots = \frac{\pi^2}{8}$. The result is

$$Q_{0,t} = u_0\,A\,H \left[ 1 - \frac{8}{\pi^2} \left( e^{-\frac{\pi^2 D t}{4\,H^2}} + \frac{1}{9}\, e^{-\frac{9\,\pi^2 D t}{4\,H^2}} + \cdots \right) \right]. \tag{39}$$

On comparing this equation with number (40) which gives the amount of material contained in the vessel it will be observed that the sum of the two at any time is $u_0 AH$, i. e., the total amount in the system, as it should be. It is also apparent from (39) that the amount of solute that escapes from an open diffusion vessel is by no means directly proportional to $D$, as has sometimes erroneously been assumed, even when $t$ and $H$ are kept constant, but that for all values of $D$, as $t$ becomes very large, $Q_{0,t}$ tends to approach the same limiting value, namely $u_0 AH$. Obviously, therefore, the amount of substance that escapes from an open vessel in a given time can be of no significance as a quantitative measure of diffusion.

The amount of solute contained at any time $t$ in a system of the sort under consideration may readily be found by multiplying equation (36) by $A\,dx$ and integrating term by term between 0 and $H$. The resulting equation is

$$Q_{0,H} = \frac{8\,u_0 A\,H}{\pi^2}\left(e^{-\frac{\pi^2 D t}{4H^2}} + \frac{1}{9} e^{-\frac{9\pi^2 D t}{4H^2}} + \ldots\right). \tag{40}$$

For calculating $Q_{0,H}$ for given values of $t$, $D$ and $H$ the brief table on p. 45 or the much more extensive one given by McKAY (1930) will be found helpful. The infinite series, in general, converges rapidly, and frequently a single term will give results of sufficient accuracy.

It is often necessary to deal with cases in which diffusion is not from the system under consideration to water, but from an external medium of some constant concentration $c$ into a system originally at a concentration of zero. This case may, for the sake of brevity, be included in the more general one in which the constant external concentration is $c$ and the initial uniform internal concentration is $u_0$; the latter concentration may of course have the particular value zero. Equations for this most general case are obtained by an obvious application of the principle of independent diffusion streams discussed above (p. 17). It is merely necessary to treat the problem as one of diffusion from a concentration of $u_0 - c$ with a basal "level" of $c$. This treatment is valid for all values of $u_0$ and $c$, but when $u_0 < c$, it is convenient to write the resulting equations in the following form

$$u = c - \frac{4\,(c - u_0)}{\pi} \sum_{p=0}^{p=\infty} \left( \frac{(-1)^p}{2p+1} \cos \frac{(2p+1)\,\pi\,x}{2H}\, e^{-\frac{(2p+1)^2 \pi^2 D t}{4H^2}} \right) \tag{41}$$

and

$$Q_{0,H} = A\,H \left[ c - \frac{8\,(c - u_0)}{\pi^2} \sum_{p=0}^{p=\infty} \left( \frac{1}{(2p+1)^2}\, e^{-\frac{(2p+1)^2 \pi^2 D t}{4H^2}} \right) \right]. \tag{42}$$

Equation (42) may be used to obtain several types of information of physiological importance. For example it is sometimes desirable to know the average concentration, $\bar{u}$, existing within a sheet of tissue of thickness $H$ exposed on one surface to the entrance of a given substance, or within a sheet of thickness 2 $H$ exposed on both surfaces to the entrance of the same substance. This value is found for both cases by dividing both sides of equation (42) by the volume $AH$, and after an obvious transformation the following very useful equation is obtained.

$$\frac{\bar{u} - u_0}{c - u_0} = 1 - \frac{8}{\pi^2}\left(e^{-\frac{\pi^2 D t}{4 H^2}} + \frac{1}{9} e^{-\frac{9 \pi^2 D t}{4 H^2}} + \ldots\right). \tag{43}$$

This equation gives the ratio of the increase of the average concentration above the basal level $u_0$ to the greatest possible increase of concentration that can occur. For the very common particular case in which $u_0$ is 0, the ratio in question represents simply the degree of saturation of the system. For calculations of degrees of saturation which are not infrequently required, a graph prepared by HILL (1928, p. 70) will be found useful.

HILL (1928, p. 68) has also constructed a table for a special case of physiological interest, namely, that of the diffusion of oxygen into muscle tissue, the value of $D$ for this case at 20° C being taken as $4.5 \times 10^{-4}$ cm.²/minute (see p. 28). Some of his values are as follows:

| $\frac{t}{100\,H^2}$ | Average Saturation |
|---|---|
| 0.3 | 0.105 |
| 0.5 | 0.169 |
| 1 | 0.239 |
| 3 | 0.414 |
| 5 | 0.534 |
| 10 | 0.732 |
| 20 | 0.911 |
| 40 | 0.990 |

Evidently, for a muscle 1 mm. thick exposed on one side, or 2 mm. thick exposed on both sides, saturation with oxygen, if diffusion alone were concerned, without any consumption of the diffusing substance by the tissue, would be 73 per cent complete in 10 minutes and 99 per cent complete in 40 minutes, etc. If the muscle were half as thick, the same values would be reached four times more rapidly; if twice as thick, four times more slowly, etc.

An interesting application of equation (43) has recently been made by ROUGHTON (1932). In earlier experiments by HARTRIDGE and ROUGHTON (1927) it was found that the time required for the uptake of a given quantity of oxygen by hemoglobin is many times as great when the hemoglobin is contained in erythrocytes as when it is merely present in solution. The question arises whether it is necessary to postulate a slowing of absorption by a relatively low permeability of the cell membrane to oxygen or whether the observed delay might be accounted for by the time required for diffusion to take place within the interior of the cells. The equation employed by ROUGHTON for the necessary calculations was the same as (43) except that partial pressures of oxygen were substituted for concentrations, and the diffusion coefficient, originally expressed in terms of pressure differences, was converted into a true diffusion coefficient by dividing it by the solubility coefficient of oxygen in the interior of the corpuscle. The latter value was taken as $4.1 \times 10^{-5}$ c.c. of $O_2$ per c.c. of corpuscle contents per mm.

of Hg, and the uncorrected diffusion coefficient as $5 \times 10^{-10}$ c.c. of $O_2$ per second per cm.² per mm. Hg per cm. The average thickness of the corpuscle, $2b$, in ROUGHTON's equation, was estimated to be $1.4 \times 10^{-4}$ cm.

For practical purposes, after the introduction of ROUGHTON's symbols, equation (43) may be simplified in two different ways. When $\frac{D}{\alpha} \frac{\pi^2}{4b^2} > 0.4$, all terms except the first may be neglected with an error lying within 0.4 per cent, whence

$$\frac{\bar{p}-p_0}{p_b-p_0} = 1 - \frac{8}{\pi^2} e^{-\frac{D}{\alpha}\frac{\pi^2}{4b^2}t}$$

where $p_b$ is the tension of oxygen in the blood, $p_0$ its initial tension in the corpuscle, and $\bar{p}$ its average tension in the corpuscle at the time $t$. When $\frac{D}{\alpha} \frac{\pi^2}{4b^2} < 0.4$, then according to INGHAM, whom ROUGHTON quotes (see also in this connection p. 102)

$$\frac{\bar{p}-p_0}{p_b-p_0} = \frac{2}{b}\sqrt{\frac{Dt}{\alpha\pi}}$$

within 0.2 per cent. By means of these two relations the values of the average saturation of the erythrocyte are calculated at different times, $p_0$ here being taken as equal to zero. Some of the values which ROUGHTON obtained in this way are the following:

| Time (seconds) | Degree of Saturation |
|---|---|
| 0.0001 | 0.554 |
| 0.0002 | 0.755 |
| 0.0003 | 0.866 |
| 0.0004 | 0.928 |
| 0.0005 | 0.960 |
| 0.0006 | 0.978 |
| 0.0007 | 0.988 |

Taking now values of $p_b$ equal to 75 mm. Hg and of $p_0$ equal to zero, respectively, it appears from the table that in 0.0004 seconds—this being a suitable time for the calculation—the degree of saturation obtainable if there were no combination of oxygen with the hemoglobin would be 92.8 per cent. The amount of gas dissolved in the corpuscle would therefore be 0.0028 c.c. per c.c. of cell, and the average internal tension, $\bar{p}$, would be 70 mm. Hg. From the earlier data of HARTRIDGE and ROUGHTON on the rate of combination of oxygen with hemoglobin in a homogeneous system, it is known that in 0.0004 seconds, at the tension in question, a degree of saturation of 18.7 per cent would be reached. Estimating the gas-combining capacity of the corpuscle as 0.4 c.c. per c.c. of cell, this amounts to 0.075 c.c. of oxygen per c.c. of cell. Since this is 25 times as great as the amount that could enter the corpuscle by diffusion in the same time, it follows that the latter process must be an important limiting factor in the rate of uptake of oxygen by the intact erythrocyte.

Another physiological application of equations of this general type has been made by INGRAHAM, LOMBARD and VISSCHER (1933), who wished to determine whether during the process of ultrafiltration there is time for the theoretical membrane equilibrium to be established before the filtrate passes from the region in which diffusion processes can be effective. They used what is essentially equation (43) and calculated by means of it the theoretical ratio of the average concentration difference produced in time $t$ over a distance $x$, to the equilibrium concentration difference over the same distance. A graph is given by them which shows the relation between the time and the distance over which diffusion equilibrium is practically (i. e., 95 per cent) complete, the diffusion coefficient of NaCl having been used as a typical value for purposes of calculation. In the same figure there is also indicated the distance which the filtrate could flow in the same time. A comparison of the two sets of values shows that the movement of the filtrate could scarcely be expected to interfere to a significant extent with the establishment of the membrane equilibrium.

An additional use of the same equation was made by WRIGHT (1934), who determined the diffusion coefficient of carbon dioxide in sheets of frog's skin by measuring by a volumetric method the amount of gas that had entered the sheet at various times as fractions of the amount that could enter it at equilibrium. The value of the diffusion coefficient obtained in this way, namely, $6.5 \times 10^{-4}$ is in fairly good agreement with that of $5.7 \times 10^{-4}$ obtained by a different method. It should be noted that $a$ in WRIGHT's equation 3 represents half the thickness of the tissue if the exposure be made on both sides and the entire thickness if the exposure be made on one side only.

## 9. One-dimensional diffusion processes in systems with two open boundaries.

This case is one of very great physiological importance, and arises so frequently in practical work that it will be treated with especial fullness. For convenience, it may be dealt with under the following subheadings: systems with two identical boundary conditions, systems with two dissimilar boundary conditions, steady states, and diffusion across thin membranes. It must be emphasized that the treatment here given will apply merely to the diffusion process as it occurs within the limits of a system for which the boundary conditions are known. Except for a few simple cases involving an entirely straightforward and uncomplicated use of partition coefficients, the assumption will generally be made that conditions at the two boundaries of the system under consideration are the same as those known to exist in the adjacent liquid media. Such an assumption seems very frequently to be justified and indeed necessary in physiological work, but it should not be forgotten that it is usually not the whole truth and may sometimes be very far from the truth. For the treatment of certain more complicated cases, see McKAY (1930, 1932a, b).

### a) Boundary conditions identical.

A typical physiological example of a system with two open boundaries and subject to two identical boundary conditions is a flat muscle, such as the sartorius of the frog, or a sheet of tissue of any kind exposed on both sides to the same well-stirred medium. Assuming homogeneity of the tissue, it is evident that in a case of this sort diffusion will occur symmetrically from the two exposed faces, and that two identical diffusion streams, which by the principle discussed above (p. 17) may be considered completely independent, will meet and cross one another within the tissue. The system as a whole will, therefore, behave exactly as if each stream had been reflected backwards from an impermeable partition half way between its two boundaries. For this reason, as has been shown in the preceding section, cases of symmetrical diffusion in systems with two open boundaries may very conveniently be dealt with by means of equations intended primarily for systems with only one open boundary. Several examples of such treatment have in fact already been given.

In order, however, to avoid any limitation to symmetrical initial distributions of the diffusing material, as well as to bring out certain additional mathematical principles of interest, there will be considered in the present section the case involving the two true boundaries of the system at $x = 0$ and $x = H$, respectively. This case is covered by equation (23) on p. 36 which may be employed for any initial distribution, $u = f(x)$. It will be sufficient by way of further illustration to deal merely with the two cases of greatest physiological interest, namely, those in which there is an initial uniform internal concentration of $u_0$ and an external constant concentration of zero or (assuming a partition coefficient of unity) an initial internal concentration of zero and a constant external concentration of $c$. The equation covering the former case is derived by substituting in equation (23) $f(\lambda) = u_0$, giving

$$u = \frac{4u_0}{\pi}\left(\sin\frac{\pi x}{H} e^{-\frac{\pi^2 Dt}{H^2}} + \frac{1}{3}\sin\frac{3\pi x}{H} e^{-\frac{9\pi^2 Dt}{H^2}} + \dots\right). \quad (44)$$

The symmetry of the system represented by this equation is obvious from the fact that the same value of $u$ is always obtained whether $x$ or $H - x$ be introduced into it, since for odd values of $n$, $\sin\frac{n\pi(H-x)}{H} = \sin\frac{n\pi x}{H}$. In other words, with the postulated initial distribution of material, the concentration must always be the same at equal distances from the two boundaries. It follows, therefore, that a system with one open and one closed boundary, might, if desired, be treated merely as a half of a system of the sort to which equation (44) is applicable. For example, for the boundary conditions: $u = 0$ when $x = 0$; $\frac{\partial u}{\partial x} = 0$ when $x = H$, an appropriate equation could be obtained from

(44) by merely changing $H$ to $2H$ and then considering only the values of $x$ lying between $x = 0$ and $x = H$. On making this substitution it is seen that the resulting equation is identical with (37), which in the preceding section was derived by an entirely different method [see in this connection ADAIR (1920)].

For the equally important case in which the initial internal concentration is zero and the constant external concentration is $c$ (remembering the qualification as to the partition coefficient) the appropriate equation is derived by the method already described (p. 50). It is

$$u = c\left[1 - \frac{4}{\pi}\left(\sin\frac{\pi x}{H}\, e^{-\frac{\pi^2 D t}{H^2}} + \frac{1}{3}\sin\frac{3\pi x}{H}\, e^{-\frac{9\pi^2 D t}{H^2}} + \ldots\right)\right]. \quad (45)$$

This equation has been used, among others, by MCBAIN (1909) und HILL (1910). Another case of possible practical importance is that in which with a constant external concentration of $c$ there is an initial uniform internal concentration of $u_0$. This case is readily dealt with, as before (p. 17), by the principle of independent diffusion streams. It obviously involves a "level" represented by a constant term which is the smaller of the two values, $c$ and $u_0$, and a process of diffusion in which the effective concentration is the difference between these two values. Or, if desired, the sum may be taken of two processes to which equations (44) and (45), respectively, are applicable.

It is frequently of physiological importance to know about the concentration of a diffusing substance in the innermost regions of a tissue exposed to it. For example, information might sometimes be desired as to the lowest concentration of oxygen that with a known external tension would at a given time be found anywhere in a flat sheet of tissue exposed to it. Assuming for simplicity that no consumption of the diffusing substance takes place—this limitation will later be removed (see p. 137)—the problem resolves itself into a calculation of $u$ for any given values of $t$ and $D$ for $x = \frac{H}{2}$. For this value of $x$, equation (45) assumes the simpler form

$$u = c\left[1 - \frac{4}{\pi}\left(e^{-\frac{\pi^2 D t}{H^2}} - \frac{1}{3}\, e^{-\frac{9\pi^2 D t}{H^2}} + \frac{1}{5}\, e^{-\frac{25\pi^2 D t}{H^2}} - \ldots\right)\right]. \quad (46)$$

An analogous relation for the maximum concentration existing at any time within the tissue is obtained from equation (44).

From equation (44) the concentration gradient for any value of $x$ is readily found by differentiation. The gradients of most practical importance are those at the two boundaries of the system, namely, those for $x = 0$ and $x = H$. Since these values are equal, though opposite in sign, they may be represented in a single equation. That derived from (44) is

$$\left(\frac{\partial u}{\partial x}\right)_{\text{boundary}} = \pm \frac{4 u_0}{H}\left(e^{-\frac{\pi^2 D t}{H^2}} + e^{-\frac{9\pi^2 D t}{H^2}} + \ldots\right). \quad (47)$$

The value of $\frac{\partial u}{\partial x}$, similarly calculated for $x = \frac{H}{2}$, is equal to zero, as could have been foreseen from the nature of the case.

The total amount of material that leaves the half of the system with the negative concentration gradient between the times $t = 0$ and $t = t$ may be found by multiplying the gradient, by $-DA\,dt$ and integrating. The amount for the entire system will be twice as great, namely,

$$Q_{0,t} = u_0 A H \left[1 - \frac{8}{\pi^2}\left(e^{-\frac{\pi^2 D t}{H^2}} + \frac{1}{9} e^{-\frac{9\pi^2 D t}{H^2}} + \dots\right)\right]. \quad (48)$$

This equation derived from (44) applies to outward diffusion; the corresponding one, derived from (45), for inward diffusion, is the same with the substitution of $c$ for $u_0$. It will be noted that in the case of the first equation when $t = 0$, $Q = 0$ $\left(\text{since } 1 + \frac{1}{9} + \frac{1}{25} + \dots = \frac{\pi^2}{8}\right)$ while for $t = \infty$ $Q = u_0 A H$; in other words, in infinite time, all the material in the system will escape. In the case of inward diffusion the final value of $Q$ will be $cAH$, that is, a condition of saturation will have been attained in which the external concentration exists throughout the entire volume. For many purposes it is of interest to know the relative amount of diffusing material in the system, that is, the ratio of the amount actually contained in it at any given time to the amount it originally contained or is capable of containing. In the case of inward diffusion this ratio may be called the average degree of saturation and is evidently

$$\frac{\bar{u}}{c} = 1 - \frac{8}{\pi^2}\left(e^{-\frac{\pi^2 D t}{H^2}} + \frac{1}{9} e^{-\frac{9\pi^2 D t}{H^2}} + \dots\right). \quad (49)$$

Equation (48) could equally well be obtained by calculating the total amount of material contained at time $t$ between the levels $x = 0$ and $x = H$ and subtracting this amount from the known initial amount of material in the system. Such a calculation involves merely the multiplication of both sides of equation (45) by $A\,dx$, followed by integration between the appropriate limits in the manner already discussed (p. 44).

The similarity of equations (48) and (49) to (39) and (43) in the preceding section will be noted. It is indeed obvious that the same degree of saturation, for example, must be attained in the same time whether the system have a thickness of $H$ and one open boundary or, in the case of a symmetrical diffusion process, a thickness of $2H$ and two open boundaries. In dealing with equations of this type, the short table on p. 45 or the much fuller one prepared by McKAY (1930) will be useful.

### b) Boundary conditions different.

The case just discussed, involving two open boundaries with identical boundary conditions, is exceeded in interest by the similar case in which the two boundary conditions are different; this case includes all processes of diffusion across membranes, of which innumerable examples occur both in the physiology of entire organisms and of single cells. In the latter case, and sometimes in the former, the membranes are so thin that important simplifications of the mathematical treatment are possible; however, it is best to begin with no limitations as to the thickness of the region in which diffusion occurs, and to treat the special case of thin membranes in a section by itself. A different sort of simplification results with membranes, or other regions, of greater thickness, after a steady state has been established; this very important special case may likewise be dealt with separately after the more general treatment has been completed.

For the most general case involving constant but dissimilar boundary conditions, we have to solve the diffusion equation for the initial condition $u = f(x)$ when $t = 0$ and for the boundary conditions $u = c_1$ when $x = 0$ and $u = c_2$ when $x = H$. It will be assumed that $c_1 > c_2$, though if the reverse were the case the same equations, by taking proper account of signs, would still be applicable. The origin for the measurement of distance may be taken at either boundary of the system, and for completeness the equations for both forms of solution will be presented, but the derivation of only the one mentioned above will be given in detail. The most general case of diffusion across a membrane involves some initial distribution of the diffusing material within the membrane. The problem may, however, be somewhat simplified by applying the principle of independent diffusion streams and dealing separately with two different processes. Thus, the molecules originally within the membrane may be thought of as leaving this region as if no others were present outside while those on the outside may be thought of as diffusing across the membrane as if its initial internal concentration were zero. The actual value of $u$, therefore, for any given values of $x$ and $t$ will be the algebraic sum of the values determined for the two streams separately. But since an equation is already available for the outwardly directed stream [number (23), p. 36], this stream may be neglected, and it is necessary merely to obtain an equation for a stream across an originally empty region. The problem to be dealt with here, therefore, is to find a solution of the diffusion equation which reduces to $c_1$ when $x = 0$, to $c_2$ when $x = H$ and to zero when $t = 0$.

Such an equation may readily be obtained by using sine-exponential terms of the form $\sin \frac{n\pi x}{H} e^{-\frac{n^2\pi^2 Dt}{H^2}}$ which become equal to zero for $x = 0$ and for $x = H$, and then by adding other terms of a different sort, which are likewise solutions of the diffusion equation (8) but which permit the boundary conditions to be satisfied. A few trials lead to an equation of the form

$$u = c_1 - \frac{c_1 - c_2}{H} x + \sum_{n=1}^{n=\infty} \left( a_n \sin \frac{n \pi x}{H} e^{-\frac{n^2 \pi^2 D t}{H^2}} \right) \tag{50}$$

which evidently satisfies both boundary conditions. It is obvious that in order that the initial condition may also be satisfied it is necessary to choose the coefficients $a_n$ so as to give rise, when $t = 0$, to a FOURIER series equal to $-c_1 + \frac{c_1 - c_2}{H} x$. This is very easy todo by finding the FOURIER sine series for unity and multiplying it by $-c_1$, then the corresponding series for $x$ and multiplying it by $\frac{c_1 - c_2}{H}$ and finally taking their sum. The values of $a_n$ so obtained, when substituted in (50) give a complete solution of the problem, for the equation now reduces to zero for $t = 0$, and the initial as well as the two boundary conditions are satisfied. The solution obtained in this way is

$$\left. \begin{aligned} u = c_1 - \frac{c_1 - c_2}{H} x - \frac{4 c_1}{\pi} \left( \sin \frac{\pi x}{H} e^{-\frac{\pi^2 D t}{H^2}} + \frac{1}{3} \sin \frac{3 \pi x}{H} e^{-\frac{9 \pi^2 D t}{H^2}} + \dots \right) + \\ + \frac{2 (c_1 - c_2)}{\pi} \left( \sin \frac{\pi x}{H} e^{-\frac{\pi^2 D t}{H^2}} - \frac{1}{2} \sin \frac{2 \pi x}{H} e^{-\frac{4 \pi^2 D t}{H^2}} + \dots \right). \end{aligned} \right\}$$

For the special case where $c_1 = c_2 = c$, equation (51) passes over into equation (45), previously obtained in a different manner. As has already been mentioned, the effect of any initial distribution within the membrane, other than a uniform concentration of zero, can be provided for by adding to equation (51) equation (23) with the substitution of the proper form of $f(x)$. For the very common and relatively simple case where $u = 0$ when $t = 0$, $u = c$ when $x = 0$, and $u = 0$ when $x = H$, the necessary equation is obtained by substituting in (51) $c_1 = c$ and $c_2 = 0$, giving

$$u = c - \frac{c x}{H} - \frac{2 c}{\pi} \left( \sin \frac{\pi x}{H} e^{-\frac{\pi^2 D t}{H^2}} + \frac{1}{2} \sin \frac{2 \pi x}{H} e^{-\frac{4 \pi^2 D t}{H^2}} + \dots \right). \tag{52}$$

It is also easy to show that the following equation replaces (52) when the origin is taken at the boundary where $u = 0$ instead of at that at which $u = c$

$$u = \frac{c x}{H} - \frac{2 c}{\pi} \left( \sin \frac{\pi x}{H} e^{-\frac{\pi^2 D t}{H^2}} - \frac{1}{2} \sin \frac{2 \pi x}{H} e^{-\frac{4 \pi^2 D t}{H^2}} + \dots \right). \tag{53}$$

Both equations (52) and (53) reduce to $u = 0$ for $t = 0$ and to $u = c$ and $u = 0$ for $x = 0$ and $x = H$, respectively, or vice-versa. In addition, when $t = \infty$, both become equations of straight lines as they should when a steady state has been attained (see p. 31).

From equations (52) and (53) other equations analogous to those already derived for amounts of material rather than concentrations may readily be obtained. Since no new principle is involved, it will be sufficient merely to give without discussion two of the most useful, namely, one for the amount of substance that leaves the membrane between the times $t = 0$ and $t = t$, which is

$$Q_{0,t} = \frac{DAct}{H} + \frac{2AHc}{\pi^2}\sum_{n=1}^{n=\infty}\left[\frac{(-1)^n}{n^2}\left(1 - e^{-\frac{n^2\pi^2 Dt}{H^2}}\right)\right] \qquad (54)$$

and one for the total amount of substance contained in the entire membrane at any time $t$, which is

$$Q_{0,H} = \frac{cAH}{2} - \frac{4cAH}{\pi^2}\left(e^{-\frac{n^2\pi^2 Dt}{H^2}} + \frac{1}{9}e^{-\frac{9\pi^2 Dt}{H^2}} + \ldots\right). \qquad (55)$$

Equation (52) has been employed by STEFAN (1878a) and by WÜSTNER (1915) and equation (53) by DAYNES (1920). The latter author extends the theory so far given to include an additional point of considerable interest and possible physiological importance. It has been mentioned that in diffusion between two aqueous solutions across a non-aqueous membrane, the true mathematical boundary conditions are not furnished by the concentrations existing in the aqueous solutions, but rather by those in the adjacent external layers of the membrane. These, in general, can be found only when the partition or solubility coefficient for the diffusing substance is known, and in the absence of this information it is usually necessary to be content with a permeability coefficient rather than with a true diffusion coefficient. In the case studied by DAYNES, however, which was the diffusion of hydrogen through a rubber membrane, it was possible from a single set of experiments to determine both the solubility coefficient and the diffusion coefficient. This was accomplished by working under conditions where the concentration of the escaping gas was kept practically at zero in a collecting vessel but where by a sufficiently delicate method its very low and increasing concentration in this vessel could be accurately measured.

In a system such as that studied by DAYNES, in which independent experiments showed (1) that FICK's law holds inside the membrane, (2) that absorption of a given gas is proportional to its partial pressure, (3) that there is no appreciable resistance at the surface of the membrane to the passage of the gas, and (4) that different gases do not impede each other's progress in the membrane, the amount of gas that would pass through unit area of the membrane from a region where its partial pressure is $p$ to one where it is practically zero, after the establishment of a steady state, would be

$$Q = \frac{DSp}{H} \qquad (56)$$

where $p$ is the partial pressure and $S$ is the absorption coefficent, which here takes the place of the partition coefficient. From observations of the steady state alone $D$ and $S$ cannot be separated from one another; they can, however, both be obtained in the following manner from measurements of the rate at which the steady state is approached.

From equation (53) the value of $\frac{\partial u}{\partial x}$ may readily be found. It is

$$\frac{Sp}{H} + \frac{2Sp}{\pi} \sum_{n=1}^{n=\infty} \left( \frac{(-1)^n}{n} \frac{n\pi}{H} \cos \frac{n\pi x}{H} e^{-\frac{n^2\pi^2 Dt}{H^2}} \right).$$

For the boundary, $x = 0$, at which gas is passing into the containing vessel, this expression assumes a much simpler form, and when this is multiplied by $D$ it gives the rate at which gas crosses unit area. This value is

$$\frac{dQ}{dt} = \frac{DSp}{H} \left[ 1 + 2 \sum_{n=1}^{n=\infty} \left( (-1)^n e^{-\frac{n^2\pi^2 Dt}{H^2}} \right) \right].$$

Now let $z$ be the low concentration, which can be continuously measured, in the collecting vessel, in which the volume associated with each unit of area of the membrane is $V$. Then $V \frac{dz}{dt} = \frac{dQ}{dt}$ and

$$z = \frac{DSp}{VH} \left[ t + 2 \sum_{n=1}^{n=\infty} \left( \frac{(-1)^n}{n^2} \frac{H^2}{D\pi^2} \left( 1 - e^{-\frac{n^2\pi^2 Dt}{H^2}} \right) \right) \right].$$

As $t$ increases, the value which $z$ approaches is

$$\frac{DSp}{VH} \left[ t + \frac{2H^2}{D\pi^2} \sum_{n=1}^{n=\infty} \frac{(-1)^n}{n^2} \right].$$

But since $1 - \frac{1}{4} + \frac{1}{9} - \ldots = \frac{\pi^2}{12}$ this may be written

$$z = \frac{DSp}{VH} \left( t - \frac{H^2}{6D} \right).$$

This is the equation of a straight line which cuts the axis $0 - t$ at a time, $H^2/6D$, which may be called the lag and may be designated by $L$. By plotting concentrations against times until the graph becomes practically a straight line the numerical value of $L$ can be obtained. But a quantity, $P$, called by DAYNES the permeability, and defined as the amount of gas which in the steady state crosses unit area in one second may also be readily measured; by equation (56) it is equal to $DSp/H$. Therefore, $PL = \frac{pH}{6} S$ and $S$ has been separated from the product $DS$ in equation (56), from which $D$ may now likewise be obtained.

### c) Steady states.

α) Rapidity of approach to the steady state.

It will be noted that when $t = \infty$, equation (51) reduces to

$$u = c_1 - \frac{c_1 - c_2}{H} x .$$

This equation is identical with equation (13) which was obtained by solving the general diffusion equation for the conditions defining a steady state, namely, $\frac{\partial u}{\partial t} = 0$; $\frac{\partial u}{\partial x} \neq 0$. Furthermore, by partial differentiation of equation (51) with respect to $x$ and to $t$, it appears that with increasing values of $t$, $\frac{\partial u}{\partial x}$ approaches a constant value which is independent of $x$, and $\frac{\partial u}{\partial t}$ approaches the value zero. It is evident, therefore, that in diffusion across any membrane between regions of unlike concentration a steady state must always be approached, though it is never actually reached in a finite time.

It is of considerable practical importance to be able to determine how closely the steady state has been approached in a given system at any given time. One reason why such information is desirable is that the steady state provides a very simple method, both theoretically and practically, for the measurement of diffusion coefficients. The chief obstacles in the way of its more extensive use for this purpose have been the long times required in ordinary diffusion systems for its practical establishment, and the supposed difficulty in deciding when to consider it as having been attained. A simple mathematical treatment of the problem, however, suggests means of overcoming both of these difficulties.

Consider equation (52) which furnishes the information necessary to calculate the concentration at any distance $x$ from one of the surfaces of a membrane of thickness $H$ across which diffusion is occurring from the constant concentration $c$ to the constant concentration 0. Following the general method of Dabrowski (1912) we first obtain from equation (52), by partial differentiation with respect to $x$, the concentration gradient at any given level. Multiplying the latter by $-D dt$ we then obtain the rate at which the solute would cross unit area, at right angles to the direction of diffusion. This rate proves to be

$$\frac{dQ}{dt} = \frac{Dc}{H}\left[1 + 2\sum_{n=1}^{n=\infty}\left(\cos\frac{n\pi x}{H} e^{-\frac{n^2\pi^2 Dt}{H^2}}\right)\right].$$

For the plane $x = 0$, i.e., the boundary of the membrane in contact with the solution this reduces to

$$\frac{dQ}{dt} = \frac{Dc}{H}\left[1 + 2\left(e^{-\frac{\pi^2 Dt}{H^2}} + e^{-\frac{4\pi^2 Dt}{H^2}} + \ldots\right)\right] \tag{57}$$

while for the plane $x = H$, i. e., the boundary in contact with the pure solvent the corresponding equation is

$$\frac{dQ}{dt} = \frac{Dc}{H}\left[1 - 2\left(e^{-\frac{\pi^2 Dt}{H^2}} - e^{-\frac{4\pi^2 Dt}{H^2}} + \ldots\right)\right]. \tag{58}$$

The difference between (57) and (58) would give the rate of accumulation of solute in the membrane, if this were desired. What concerns us here, however, is the closeness of the approach of the diffusion process at any given time to the theoretical steady state. For this state, obviously,

$$\frac{dQ}{dt} = \frac{Dc}{H}.$$

As a measure of the divergence of the system from its final state we may use the difference between the limiting rate and the actual rate, divided by the former, to give a fraction whose value will approach zero with increasing time. The measure of divergence, which may be designated by $\varepsilon$, therefore, has the value, for the surface at which material is leaving the membrane

$$\varepsilon = 2\left(e^{-\frac{\pi^2 Dt}{H^2}} - e^{-\frac{4\pi^2 Dt}{H^2}} + \ldots\right).$$

When $\frac{Dt}{H^2}$ is fairly large, as it must be when the steady state is closely approached, all but the first exponential term may be neglected, and we obtain the relation

$$t = \frac{H^2}{\pi^2 D} \ln \frac{2}{\varepsilon}. \tag{59}$$

The time, therefore, required to reach a rate of escape of material differing from the theoretical limiting rate by a fraction, $\varepsilon$, of the latter is seen to be inversely proportional to the diffusion coefficient and directly proportional to the square of the thickness of the membrane but not to depend in any way on the concentration of the solution from which diffusion is taking place. (See also in this connection DOMBROWSKY 1925.)

| $Dt/H^2$ | $\varepsilon$ |
|---|---|
| 0.3 | 0.10354 |
| 0.4 | 0.03860 |
| 0.5 | 0.01438 |
| 0.6 | 0.00536 |
| 0.7 | 0.00200 |
| 0.8 | 0.00072 |
| 0.9 | 0.0028 |
| 1.0 | 0.00010 |

It is instructive to tabulate certain values of $\varepsilon$ which show the divergence of the system from the steady state for different values of $Dt/H^2$.

It is even more instructive, however, to calculate from equation (59) the times required to approach with some given degree of closeness the steady state for a fixed value of $D$ and different values of $H$. For convenience, $D$ may be so chosen that $\pi^2 D = 10$; it happens that the value of $D$ which satisfies this relation is approximately that of the diffusion coefficient for NaCl or urea at or slightly below $20^0$ C, so the hypothetical case is an entirely typical one.

Values of $t$ for different values of $H$ and $\varepsilon$ for $\pi^2 D = 10$.

| $H$ \ $\varepsilon$ | 0.01 | 0.001 | 0.0001 |
|---|---|---|---|
| 10 cm. | 52.98 days | 76.01 days | 99.04 days |
| 1 cm. | 12.72 hrs. | 18.24 hrs. | 23.76 hrs. |
| 1 mm. | 7.6 min. | 10.9 min. | 14.25 min. |
| 100 $\mu$ | 4.56 sec. | 6.54 sec. | 8.55 sec. |
| 10 $\mu$ | 0.046 sec. | 0.065 sec. | 0.086 sec. |
| 1 $\mu$ | $4.6 \times 10^{-4}$ sec. | $6.5 \times 10^{-4}$ sec. | $8.6 \times 10^{-4}$ sec. |
| 0.1 $\mu$ | $4.6 \times 10^{-6}$ sec. | $6{,}5 \times 10^{-6}$ sec. | $8.6 \times 10^{-6}$ sec. |

It will be seen from this table that with the diffusion coefficient in question, it would require 53 days to approach the limiting rate of escape within one per cent, if the distance of diffusion were 10 cm. Evidently, therefore, steady state experiments are impracticable in systems of this order of magnitude. The case is very different, however, for smaller distances. In a system having a thickness of 1 mm., for example, which considerably exceeds the thickness of the membranes used by NORTHROP and ANSON (1929), a few minutes would suffice for the attainment of a practically constant rate of escape of the diffusing substance. Steady state methods may therefore be employed with entire confidence with systems of these dimensions. In the case of membranes of the sort encountered in living cells, where a thickness of 1 $\mu$ may be considered extraordinarily great, and one of less than 0.1 $\mu$ not unusual, steady states under otherwise appropriate conditions, could be established practically instantaneously.

$\beta$) Practical applications of steady state methods.

Steady states have played an important part historically in the study of diffusion processes. The earliest experimental test of FICK's law, made at the time of its formulation (FICK 1855), was carried out, after attempts to use a more general method had failed, by measurements in a system in which a steady state had been established. FICK's procedure was as follows: A diffusion vessel of constant cross-section was placed in communication at its lower end with a reservoir containing a saturated salt solution, whose concentration was maintained by the presence of salt crystals. The upper open end of the diffusion vessel lay below the surface of a body of water contained in a larger outer vessel. Because of its greater specific gravity, the diffusing solution left the upper end of the diffusion vessel immediately on escaping from it, thereby maintaining at this end of the vessel a concentration of zero. After the practical attainment of the steady state, which could be hastened by filling the diffusion vessel with layers of salt solution of decreasing density instead of with water or with a solution of uniform concentration, FICK determined in situ by means of a glass bead suspended from the beam of a balance the specific gravity of the solution at different levels in the diffusion vessel. Doubtless the rather poor agreement of his observations with the theory was due to the unavoidable convection currents set up by this somewhat crude method of measurement. The results he obtained did, nevertheless, point clearly to a linear relation between concentration and distance from the bottom of the vessel, as was demanded by such an equation as (13). Because of the historical importance of these results, the following partial list of figures obtained by FICK may be presented.

| Distance below the surface of the diffusion vessel in mm. | Excess of specific gravity over that of water. |
|---|---|
| 10 | 0.009 |
| 32.2 | 0.032 |
| 54.4 | 0.053 |
| 76.6 | 0.073 |
| 98.8 | 0.093 |
| 121.0 | 0.115 |
| 143.2 | 0.135 |
| 165.4 | 0.152 |

Equation (14) suggests a second method by which the steady state method may be used to test FICK's theory. Since the rate of escape of the solute in the system just described must, by the theory in question, be proportional to the concentration gradient, $c/H$, it follows that it must be exactly inversely proportional to the length of the diffusion vessel. Experiments by FICK with vessels of three different lengths showed that this relation was, in fact, obtained. FICK determined from his steady state experiments a diffusion coefficient for NaCl, which he expressed in units different from those now employed; his figures, however, as recalculated by STEFAN (1878b) in spite of the crudity of his methods, are of the right order of magnitude.

Following FICK, other workers have used the steady state method for studying free diffusion processes. One of the first was STEFAN (1878a) who employed it for measuring the diffusion coefficient of carbon dioxide in water and in alcohol. The principle of the method was to allow the gas to diffuse from a known tension in the closed end of a capillary tube through a freely movable layer of water in the same tube to a constantly maintained tension of practically zero (ordinary air). By measuring the rate of movement of the layer of water and making the necessary corrections for the absorption of air, which is, however, slow in comparison with that of carbon dioxide, the rate of diffusion of the gas in the liquid is obtained, and from it the diffusion coefficient may be calculated. STEFAN's experiments showed clearly and quantitatively that the rate of diffusion is inversely proportional to the thickness of the layer in which the steady state of diffusion exists, and his value of $D$, obtained in this way, is in good agreement with that which he and other workers found by other methods.

An extensive and very careful set of determinations of diffusion coefficients of various salts has been made by CLACK (1908—1924) chiefly by the steady state method. An important innovation introduced by this investigator is the measurement of the rate of escape of the solute from the diffusion vessel into a surrounding body of water, not by chemical analysis of the latter, as in FICK's experiments, but by continuous observations of the decrease in the specific gravity of the diffusion vessel as salt escapes. For an accurate determination of diffusion coefficients by this method it is necessary to take into account, as CLACK has done, the slow inflow of water into the diffusion vessel which makes good the loss of volume caused by the escape of salt. This effect, which is not provided for in FICK's law, has been investigated in detail by GRIFFITHS (1898, 1899, 1916a, b) who has even made it the basis of a unique method for determining diffusion coefficients. In general, the results obtained by this last method, by a steady state method involving chemical analyses (GRIFFITHS, DICKSON and GRIFFITHS 1916), and by CLACK's specific gravity method are in excellent agreement. One great advantage of the steady state method is that in CLACK's hands it has permitted the determination of definite diffusion coefficients for fixed concentrations; this is not usually possible with the methods commonly employed, which merely give a sort of average value for the entire range of concentrations present in the region of diffusion.

In recent years, interest in the steady state method has shifted from cases of free diffusion to cases of diffusion across membranes of a particular type, which permit ordinary diffusion coefficients for water to be obtained indirectly. The membranes in question have a coarsely porous structure, and while they are able to prevent convection currents between two stirred solutions which they separate, they offer little opposition to the diffusion of even large molecules. It had very early been shown by STEFAN (1878a) that convection currents, which had proved so troublesome in the work of WROBLEWSKI (1877, 1878), can almost be prevented by allowing diffusion to take place in capillary tubes 1 mm. or less in diameter. DABROWSKI (1912) took advantage of this fact to construct artificial membranes of known thickness and known cross-section with respect to their aqueous channels by cementing together large numbers of capillary glass tubes. Such membranes furnish an excellent connecting link between cases of free diffusion and those of diffusion across ordinary membranes. Since the total area available for diffusion—which is however considerably less than the total area of the membrane—can be measured directly, the ordinary diffusion equations can be used with them after the substitution of the proper calculated value of $A$. It should be noted, however, that this treatment is justified only if the solutions between which diffusion is occurring are stirred, since the rate of diffusion through sufficiently separated capillary tubes into an unstirred region is proportional not to the total cross-section of the tubes but to the sum of their diameters. This principle, incidentally, is one of much importance in plant physiology in connection with the functions of the stomata of the leaves of the higher plants (Brown and Escombe 1900, Brown 1901, 1918).

Though, in theory, the type of membrane used by Dabrowski leaves little to be desired, since its exact structure is known, it has not proved to be a very practical one for actual experimental purposes. Various workers have therefore used other porous membranes of unknown structure and have then standardized them in some way in order that conclusions might be drawn about diffusion in their aqueous portions. Hüfner (1897, 1898), for example, used thin sheets of the mineral hydrophane, which, after the absorption of water, contains aqueous channels available for diffusion. By methods which he describes in his original paper, Hüfner was able to find a water equivalent for a given thickness of hydrophane and so to obtain diffusion coefficients for water alone. Somewhat later Jablczynski (1909) used as a porous membrane ordinary cigarette paper and standardized it by measuring the rate of diffusion through it of some substance whose diffusion coefficient is already known.

Recently, a very useful method has been introduced by Northrop and Anson (1929), and this method is at present one of the most satisfactory of all those employed for the measurement of diffusion coefficients. It originally consisted in the measurement of the constant rate of diffusion, after the attainment of the steady state, between a stirred solution of the diffusing substance and a stirred body of water through a thin membrane of sintered glass or alundum. This membrane is sufficiently porous to permit the free passage of molecules of proteins, enzymes, etc., but it completely eliminates convection currents. While the total cross-section and other characteristics of its pores cannot be determined by direct observation, the membrane may be standardized by measuring the rate of diffusion through it of some substance such as HCl whose diffusion coefficient in water is accurately known. In this way a constant for any particular membrane can be obtained, which, when applied to the passage through it of some other substance, permits the determination of the diffusion coefficient of the latter. By means of this method, Northrop and Anson obtained the value of the diffusion coefficient of hemoglobin already mentioned above (p. 13).

The same method has since been employed to very good advantage in studies on the diffusion of such substances as the following: pepsin (Northrop 1930), trypsin (Northrop and Kunitz 1932, Scherp 1933), hydrogen peroxide (Stern 1933), catalase (Zeile 1933). The method has also been adapted to a variety of important problems by McBain and his co-workers (1931—1934). One modification of the original steady state method made by McBain is that after an approximately linear concentration gradient has been attained across the membrane no attempt is made to maintain a constant concentration difference between the two sides of the membrane, but an appropriate equation similar to one derived but not used by NORTHROP and ANSON (1929) is employed for the calculation of $D$ from the observed concentration changes. A theoretical justification for this procedure has recently been furnished by BARNES (1934).

The steady state method of determining diffusion coefficients is of particular importance in physiology since it was by means of this method that KROGH (1919a) obtained the diffusion coefficients of a number of physiologically important gases through various tissues. It will be noted that in KROGH's experiments what was desired was not the diffusion coefficient of a given gas in water, using a membrane merely as a convenient means of eliminating convection currents, but rather the diffusion coefficient of the gas in the membrane itself. The method employed was to separate two chambers by a membrane of as uniform thickness as possible. In one chamber, the gas was kept at a known tension, either as a gas at a known partial pressure or in a solution previously brought into equilibrium with the gas at this pressure, which, as experience showed, leads to the same result. In the other chamber, the tension was kept approximately at zero; in the case of $CO_2$ by absorption with NaOH and in the case of $O_2$ and CO by absorption by means of hemoglobin. The rate of diffusion under these conditions was then obtained after the establishment of a steady state, which, with the thin membranes employed, must have been attained very quickly. In the following table are given the values obtained with several animal tissues and other materials by KROGH for oxygen at 20° C. (The unit of pressure difference is here taken as 1 atmosphere per micron of thickness of the tissue.) KROGH's constant for carbon dioxide is approximately 35 times as great as that for oxygen. The steady state method has also recently been used by WRIGHT (1934) to determine the diffusion coefficient of carbon dioxide in frog's skin. His value is of the order of magnitude of that found by KROGH.

| | |
|---|---|
| Water | 0.34 (HÜFNER) |
| Gelatin | 0.28 |
| Muscle | 0.14 |
| Connective tissue | 0.115 |
| Chitin | 0.013 |
| Rubber | 0.077 |

γ) Steady states in systems of varying cross section.

It has been mentioned above that the solution of certain diffusion problems, whose general mathematical treatment would be difficult, becomes easy when a steady state may be assumed to exist. One such problem is that of diffusion in a vessel of variable cross-section. It arose very early; in fact, it appeared in the original experimental tests which FICK (1855) made of his theory. As an illustration of the principles involved, this particular case may be discussed with a somewhat fuller explanation of the successive mathematical steps than was given by FICK. In the experiments in question, the situation may be described (using as far as possible the symbols already employed) as follows. At the small end of a funnel-shaped vessel, the radius of the vessel here being $r_1$, a concentration of $c$ is constantly maintained. At the large end of the vessel where the radius is $r_2$ and the distance is $H$ from the small end, the concentration of the diffusing substance is kept at zero by the difference in the specific gravities of the diffusing solution and pure water. After a steady state has been established, it is required to find the concentration at any level of the vessel, and the constant rate at which the solute leaves the vessel.

Consider an elementary volume of thickness $dx$ at right angles to the direction of diffusion. The rates both of the entrance of the diffusing substance into, and of its escape from, this element are governed by FICK's law. However, not only does the concentration gradient $\frac{\partial u}{\partial x}$ change with distance but the cross-section of the vessel changes as well. The rate of entrance into the elementary volume, as before, will be $-DA\frac{\partial u}{\partial x}$. The rate of escape, however, will be

$$-D\left(A+\frac{dA}{dx}dx\right)\left(\frac{\partial u}{\partial x}+\frac{\partial^2 u}{\partial x^2}dx\right).$$

The difference between the two rates, i. e., the rate of accumulation, is obtained by subtracting the rate of escape (after dropping infinitesimals of a higher order than the first) from the rate of entrance. The rate of accumulation may also be expressed as $\frac{\partial u}{\partial t}$ multiplied by the volume of the element. Again dropping infinitesimals of higher order, we obtain an expression which, when equated to the first, after the removal of the common factor $dx$, gives the desired equation, namely

$$\frac{\partial u}{\partial t}=D\left(\frac{\partial^2 u}{\partial x^2}+\frac{1}{A}\frac{\partial u}{\partial x}\cdot\frac{dA}{dx}\right). \qquad (60)$$

Equation (8) is evidently a special case of equation (60) in which $\frac{dA}{dx}$ is equal to zero. It will be noted that the notation of partial differentiation is not applied to $A$, since this variable does not change with time but only with distance.

Equation (60) may now be applied to the funnel-shaped vessel by imagining the latter to be extended beyond its smaller end until its radius has become zero, i. e. until a complete cone has been produced; $x$ is then measured from the tip of the cone in the direction in which diffusion is occurring. Let the distance of the small end of the funnel from the origin be $h$ and that of the large end $h+H$; $H$, as previously, representing the distance through which diffusion actually occurs. Let the concentration at the small end of the funnel be $c$ and that at the large end be 0. Represent by $m$ the ratio $r/x$, i. e. $r=mx$, where

$m$ is the tangent of one-half the angle between opposite sides of the vessel. We have, therefore

$$A = \pi m^2 x^2 \quad \text{and} \quad \frac{dA}{dx} = 2 \pi m^2 x.$$

Substituting these values in equation (60) and remembering that for the present case $\frac{\partial u}{\partial t} = 0$ we obtain the differential equation for the steady state of diffusion in a funnel

$$\frac{d^2 u}{d x^2} + \frac{2}{x} \frac{d u}{d x} = 0.$$

This equation may readily be solved by introducing a new variable $p = \frac{du}{dx}$. The resulting equation

$$\frac{dp}{dx} + \frac{2}{x} p = 0$$

is linear and of the first order and its solution, obtained by using the integrating factor $x^2$, is

$$p = \frac{C_1}{x^2}.$$

Next substituting for $p$ its value $\frac{du}{dx}$ and again integrating we obtain

$$u = C_2 - \frac{C_1}{x}.$$

The constants of integration in this equation are evaluated from the information that when $x = h$, $u = c$ and when $x = H + h$, $u = 0$. Introducing the proper values of the constants the equation becomes

$$u = \frac{h(H+h)c}{Hx} - \frac{hc}{H}. \tag{61}$$

Equation (61) may be used to find the concentration corresponding to any value of $x$ lying between $x = h$ and $x = h + H$. It was doubtless by means of an equation such as this, though details are not given, that FICK calculated the theoretical concentrations at different levels, which on comparison with those which he observed by the specific gravity method showed a satisfactory agreement, thereby giving support to his theory.

The amount of substance that in a case of this sort would leave the vessel, or cross any plane at right angles to the direction of diffusion, in a unit of time is obtained by differentiating to find $\frac{\partial u}{\partial x}$ and then multiplying by $-DA$. Remembering that $A = \pi m^2 x^2$ we obtain in this way

$$\frac{dQ}{dt} = \frac{D \pi m^2 c h (H+h)}{H}. \tag{62}$$

But $mh = r_1$ and $m(H+h) = r_2$; equation (62) may therefore be written in its most useful form which applies to finite values of $Q$ and $t$

$$Q_{0,t} = \frac{D \pi c r_1 r_2 t}{H}. \tag{63}$$

It is apparent from equation (63) that just as in a steady state in a vessel of constant cross-section, so here, the amount of substance that crosses any given level in unit time is directly proportional to $D$ and to $c$ and inversely proportional to $H$. The close relation of the two cases is seen by placing $r_1$ equal to $r_2$, when equation (63) becomes identical with (14). A further point of interest about equation (63) is that it shows that the rate of diffusion would be exactly the same in the steady state if the large and the small ends of the funnel were interchanged. Finally, by a simple application of the principle of maxima and minima, it is easy to show that for the same average radius of the vessel, i. e. for any fixed value of $r_1 + r_2$ the rate of diffusion is greatest when $r_1 = r_2$.

### δ) Steady states of diffusion across two dissimilar layers.

Another problem of some practical importance is the following. Suppose that diffusion occurs, not through a single homogeneous medium in which the diffusion coefficient is constant, but first through a layer of one substance, of thickness $h_1$ in which the diffusion coefficient is $D_1$ and immediately thereafter through a layer of some other substance of thickness $h_2$ in which the diffusion coefficient is $D_2$. Cases of this sort are frequently encountered in physiology, diffusion occurring through membranes composed of two dissimilar layers such as, for example, frog skin. A similar principle might have been used in the standardization of the hydrophane membranes employed by HÜFNER (1897, 1898) for measuring the diffusion coefficients of gases. Provided that a steady state may be assumed to exist, all such cases can readily be dealt with as follows. (Partition effects are here neglected but could readily be introduced if desired.)

For simplicity, assume that the cross-section of the system is unity. Let the free boundary of the first substance, for which the symbols $h_1$ and $D_1$ are employed, be maintained at the concentration $a$, and the free boundary of the other substance at the concentration $b$ $(a > b)$. For a steady state, the gradient within each medium must be uniform, but will in general be different for the two media; the gradients in question may be represented by $g_1$ and $g_2$, respectively. Furthermore, since in a steady state the rate of movement of the solute is everywhere the same, we have by FICK's law

$$D_1 g_1 = D_2 g_2 \quad \text{or} \quad g_1 = \frac{D_2}{D_1} g_2 .$$

But

$$h_1 g_1 + h_2 g_2 = b - a .$$

Therefore

$$-g_2 = \frac{D_1 (a-b)}{D_1 h_2 + D_2 h_1} \quad \text{and} \quad -g_1 = \frac{D_2 (a-b)}{D_1 h_2 + D_2 h_1} .$$

From either value of $g$ we obtain the rate at which the solute would pass through the two-layered system, namely,

$$\frac{dQ}{dt} = \frac{D_1 D_2}{D_1 h_2 + D_2 h_1} (a - b). \tag{64}$$

When $D_1 = D_2$ equation (64), as it should, reduces to the simple form already given above [(14), p. 31)].

To find the concentration at any level in the first layer at a distance $x$ from its outer boundary it is necessary merely to use the relation $u = a + g_1 x$ or

$$u = a - \frac{(a-b) D_2 x}{D_1 h_2 + D_2 h_1}. \tag{65 a}$$

Similarly at any distance $x$ from the outer boundary of the second layer the concentration is $b - g_2 x$ or

$$u = b + \frac{(a-b) D_1 x}{D_1 h_2 + D_2 h_1}. \tag{65 b}$$

If desired, more complicated cases of this sort may be dealt with in a similar manner. Fürth (1927c) has discussed the most general possible case involving an indefinitely large number of layers.

### ε) Steady states and variable diffusion coefficients.

Though the diffusion coefficient appearing in Fick's law has so far been treated as if it were a constant, it is in reality known to show a considerable variation with changes in the concentration of the solutions used in its determination as may readily be seen by consulting any extensive tables such as the Landolt-Börnstein Tabellen or the International Critical Tables. While the values published in these tables demonstrate clearly the variability of the diffusion coefficient with changing concentrations, they do not give an exact idea of its value at any single concentration, since when diffusion is allowed to occur into water from some given initial concentration, the distribution of the diffusing material at any time must depend in a complicated way on simultaneous diffusion processes involving all concentrations between the initial concentration and zero. The value of $D$ obtained from experiments of this sort is therefore really a sort of composite or average value, and not one which is precisely defined for a single concentration. Only by making the difference between the initial layer of solution and the remainder of the system extremely small could a truly definite value of $D$ be obtained from experiments of the usual sort, but the practical difficulties of working with systems of this sort would be very great.

By the steady state method, however, it is possible on the assumption that the rate of diffusion is directly proportional to the concentration gradient, and that the constant of proportionality varies with the concentration, to calculate precise values of $D$ for definite concentrations. This has been done by CLACK (1914, 1916, 1921) whose first approximate treatment of the subject, with the omission of his later allowances for the convection currents caused by volume changes, will suffice to illustrate the principles involved. As before, to avoid confusion, the symbols employed in the present paper will be used rather than those of CLACK.

Suppose that the amount of material, $Q$, that in unit time crosses a region of unit area and of thickness $H$ under the conditions of a steady state be measured. Let the fixed concentrations between which passage of material occurs be $c$ and 0, respectively, and let the distance $x$ be measured from the boundary where the concentration is zero. On the assumption of a single diffusion coefficient and a uniform concentration gradient we should have the relation for unit time

$$Q = {}_0D_c \frac{c}{H}$$

where, following CLACK, the symbol ${}_0D_c$ is used to represent the theoretical diffusion coefficient calculated in the usual manner on these assumptions. In reality, in the actual system we have to do with a series of different diffusion coefficients for all concentrations from $c$ down to zero. Let the true diffusion coefficient for any concentration $u$ be represented by $D_u$. Then, since in a steady state the rate of passage of material must everywhere be the same, and must obey FICK's law, the concentration gradient cannot be uniform but must vary with $D_u$ in the following manner

$$Q = D_u \frac{d\,u}{d\,x}$$

where $Q$ has the same value as before. From this equation it follows that

$$D_u = Q \frac{d\,x}{d\,u}.$$

Suppose now that different values of ${}_0D_c$ be determined experimentally for a series of decreasing values of $c$ and let the results be plotted in the form of a curve. From this curve, by drawing tangents, the value of $\frac{d\,({}_0D_c)}{d\,c}$ can be determined for any value of $c$. But each of the values of $c$ so chosen exists in the system first studied as the value of $u$ corresponding to some particular value of $x$. Since $Q$ is constant we may write

$$Q = {}_0D_c \frac{u}{x} \qquad \text{or} \qquad Q\,x = {}_0D_c\,u$$

where $u$ has one of the chosen values of $c$. Therefore, after differentiating with respect to $c$ and substituting the value of $D_u$ given above we obtain finally

$$D_u = {}_0D_c + c\,\frac{d\,({}_0D_c)}{d\,c}. \tag{66}$$

From this relation $D_u$ may readily be obtained from the observed data. The following are typical values of $D_u$ calculated in this way for KCl at 18.5° C. (CLACK 1914).

| Concentration | $D_u$ |
|---|---|
| 0.05 | 1.388 |
| 0.10 | 1.430 |
| 0.20 | 1.467 |
| 0.40 | 1.493 |
| 0.60 | 1.504 |
| 0.80 | 1.515 |
| 1.00 | 1.527 |
| 1.50 | 1.555 |
| 2.00 | 1.584 |

### d) Thin membranes.

Inasmuch as the subject of membranes has recently been dealt with in this journal (KRIJGSMAN 1932), no attempts will be made here to enter into the more general aspects of the question, involving as they do an enormous literature in such important fields as the structural, electrical and other physical properties of membranes, theories of membrane and cell permeability, osmotic phenomena, membrane equilibria, etc. The present discussion will be confined to certain mathematical aspects of diffusion across regions of relatively small thickness, which may for convenience be called membranes, but which will for the most part be considered to be homogeneous in structure and to be governed by known boundary conditions. In systems which may with sufficient accuracy be assumed to possess these characteristics the necessary mathematical treatment becomes greatly simplified and certain new and otherwise difficult problems may be attacked with success.

#### α) Diffusion of a solute alone.

By the methods discussed in the preceding sections, the course of diffusion of a solute across membranes of any thickness can readily be dealt with when the two boundary conditions remain constant. When, however, one or both of these conditions are subject to change with time, the problem becomes far more difficult, and an elementary treatment of it is usually impracticable except in the case now to be considered, in which the membrane is assumed to be very thin. Fortunately, many biological membranes are of this character; indeed the membranes occurring in single cells, from the point of view of diffusion processes, are always so. It has already been shown that in membranes whose thickness is of the order of magnitude of one micron or less a steady state may for all practical purposes by considered to be established almost instantly. Under these circumstances, the non-constancy of the boundary conditions presents no mathematical difficulties.

After a steady state has been established across a membrane, the diffusion gradient within the membrane is everywhere the same, namely

$$\frac{du}{dx} = -\frac{c_1 - c_2}{H} \tag{67}$$

where $c_1 - c_2$ is the concentration difference between the two sides of the membrane and $H$ is its thickness. If, now, the value of $c_1 - c_2$

be changed, a new steady state could again be approached at a rate that increases rapidly with the thinness of the membrane. With a very thin membrane it is reasonable to suppose that even if $c_1 - c_2$ were changed continuously, there would at all times be maintained an almost linear fall of concentration across the membrane, represented at least as a close approximation by equation (67). That this is, in fact, the case, even for considerably thicker membranes than those here under consideration, has recently been shown by BARNES (1934) by a mathematical analysis too complicated for reproduction here. Advantage of this fact has been taken by McBAIN and his collaborators (see p. 76), who after establishing an initial linear fall of concentration across membranes of alundum and sintered glass similar to those first used for diffusion studies by NORTHROP and ANSON (1929) have then permitted the gradient across the membrane to change with the further progress of diffusion.

When the thinness of a membrane permits the assumption of a single linear fall of concentration across the region of diffusion, the way is opened for a simple treatment of a great variety of diffusion problems. One of the most useful is the following. Suppose that two bodies of stirred liquid of volumes $V_1$ and $V_2$ respectively, be separated by a thin membrane of thickness $H$. Let the amounts of the diffusing solute initially present in the two volumes, in the order above mentioned, be $a$ and $b$ (where $a > b$) and let $Q$ be the amount of solute that has at the time $t$ passed from $V_1$ to $V_2$. By FICK's law, for a system of this sort

$$\frac{dQ}{dt} = -DA\frac{c_2 - c_1}{H} = \frac{DA}{H}\left(\frac{a-Q}{V_1} - \frac{b+Q}{V_2}\right).$$

This equation when simplified and integrated (the integration constant being evaluated from the information that $Q = 0$ when $t = 0$) becomes

$$\frac{DAt}{H} = \frac{V_1 V_2}{V_1 + V_2} \ln \frac{aV_2 - bV_1}{aV_2 - bV_1 - (V_1 + V_2)Q}. \tag{68}$$

This with the use of slightly different symbols is the same equation as that derived by NORTHROP and ANSON (1929).

It is frequently convenient, for any given membrane, to combine $D$, $H$, and $A$ and sometimes a partition coefficient, which may or may not be known, to form a single permeability constant, $K$; so long as the same membrane is used this constant suffices to define the relative rates of diffusion across it of different substances; though to obtain absolute values of these rates, or even to obtain the same numerical values with different membranes, it is necessary to employ true diffusion constants. In the case of cell membranes, for which $A$ can be measured but for which, in general, $H$ cannot, and in which the role of partition factors is more or less obscure, it is advantageous to incorporate $D$, $H$ and if necessary, $S$, into a single constant while keeping $A$

separate. A permeability constant of this type, while of no value for measuring physical diffusion rates as such, may be of great usefulness for comparing the permeabilities of different cells to different substances under otherwise comparable conditions. Such a constant, here represented for a penetrating solute by $k_1$, is a numerical measure of the amount of material that in unit time would cross unit area of the membrane, not with a concentration gradient of unity across the membrane, but with a unit difference in concentration on the two sides of the particular membrane. If, as is frequently the case, the concentrations measured are those in aqueous solutions adjacent to the membrane, and passage across the membrane involves solution in some non-aqueous phase, a knowledge of the partition coefficient of the diffusing substance between the two phases as well as of the thickness of the membrane is needed to obtain the true diffusion coefficient from the permeability coefficient. If, on the other hand, diffusion is through pores, then a numerical factor is needed to convert the over-all surface of the membrane into the surface actually available for diffusion.

It is fortunate that in the very cases in which the thickness of membranes is most difficult to measure with accuracy, and in which it is almost impossible to determine the extent to which partition factors are involved in the diffusion process, namely, in studies on single living cells, it is usually more important to know how permeable different cells may be under the same conditions, or how permeable the same cell may be under different conditions, than the exact physical reason why in a given experiment some particular degree of permeability is found. Though it would be most interesting to be able to analyze cell permeability into separate factors having to do with structure, thickness, solubility, etc., it must be frankly admitted that such an analysis is not likely in most cases to be possible in the near future; in the meantime, valuable information may be accumulated concerning the actual magnitude of the exchange between the cell and its surroundings under properly standardized conditions and for this purpose the permeability constant, as defined above, is most useful.

Equation (68) assumes a simpler form in special cases. Thus, when $b = 0$, i. e., when solute diffuses into a region originally filled with water, we have

$$K t = \frac{V_1 V_2}{V_1 + V_2} \ln \frac{a V_2}{a V_2 - (V_1 + V_2) Q} \tag{69}$$

and with the further simplification that $V_1 = V_2 = 1$

$$K t = \frac{1}{2} \ln \frac{a}{a - 2Q}. \tag{70}$$

This is essentially the equation used by LUNDSGAARD and HOBØLL (1926) for quantitative measurements of the permeability of collodion membranes; for a given solute such as dextrose, these investigators found over considerable periods of time a very satisfactory constancy of $\ln \frac{a}{a-2Q}$ indicating the applicability of equation (70) to the case in question. JABLCZYNSKI (1909), BROOKS (1925) and McBAIN and LIU (1931) have also used the same equation.

A very important case, which has received extensive application in physiological and chemical work is that in which the concentration on one side of the membrane is maintained at a constant value, while it is allowed to vary on the other. For such a case, according to FICK's law

$$\frac{dQ}{dt} = \frac{DA}{H}(u - c) \tag{71}$$

where $u$ is the variable concentration of the solution from which diffusion is occurring whose volume, $V$, remains constant, $c$ the constant external concentration and $Q$ the amount of solute that crosses the membrane.

One important application of this equation is found in connection with the quantitative measurement of the permeability of cells to diffusing solutes. An early suggestion that an equation of this type be employed for the purpose was made by RUNNSTRÖM (1911) and was later carried out by BÄRLUND (1929). Certain qualifications as to the applicability of the equations of RUNNSTRÖM and BÄRLUND are discussed by JACOBS (1933a, p. 429). The most extensive and satisfactory use yet made of an equation of this type to the problem of cell permeability appears in a recent paper by COLLANDER and BÄRLUND (1933) who worked with cells of the plant *Chara*. This investigation indicates very clearly the applicability of FICK's law to phenomena of cell permeability, and serves at the same time to justify certain simplifying assumptions commonly made in all mathematical treatments of this subject. The most important of these assumptions in the present case is that the delay in diffusion is very slight in other regions than across the cell membrane and that therefore the solute may be considered at any time to be uniformly distributed throughout the entire cell volume.

Another set of physiological phenomena to which equation (71) has been applied have to do with the diffusion of gases from an enclosed space into the body of a large organism. For these applications, concentrations rather than amounts will be used and $Q$ will therefore be expressed in terms of $u$. Representing the presumably constant volume of the space from which diffusion occurs by $V$ and writing $-V\,du$ in place of $dQ$, we have

$$-\frac{du}{dt} = \frac{DA}{VH}(u - c) = K(u - c).$$

On integration, remembering that when $t = 0$, $u = u_0$, we obtain

$$-Kt = \ln\frac{u-c}{u_0-c} \quad \text{or} \quad u = c + (u_0 - c)\,e^{-Kt}. \tag{72a}$$

If $c$ be greater than $u_0$ and the diffusion, therefore, be in the reverse direction, the same equation is written

$$u = c - (c - u_0)\,e^{-Kt}. \tag{72b}$$

Equations of this type, though derived and expressed in a somewhat different manner, were found by McIver, Redfield and Benedict (1926) to describe very satisfactorily the process of diffusion in either direction of various gases between the stomach and intestine of the dog and the circulating blood. Not only did diffusion follow the exponential law indicated in the equations, but the process was found to have an actual velocity of the order of magnitude to be expected from the measured values of $V$ and $A$ together with Krogh's value of the diffusion coefficient for a similar tissue and a reasonable estimate of $H$, which cannot be measured very exactly.

If in equations (72a) and (72b) $c$ be given the value zero, they assume a simpler form. Expressed logarithmically

$$K = \frac{\ln u_1 - \ln u_2}{t_2 - t_1}. \tag{73}$$

This is the equation used by Krogh and Krogh (1910) in studying the absorption from the lungs of a gas (carbon monoxide) whose tension in the blood could be considered to remain practically at zero throughout the experiment. By measuring at two different times the tensions of this gas in the lungs of a human subject who had inhaled a small quantity, $k$ could readily be calculated. Since, in the case of the human lungs, it is very difficult to measure accurately any of separate quantities that enter into the permeability constant, this constant can be said merely to define the amount of the gas in question that would enter the lungs of a given individual under the conditions of the experiment, in unit time, if the tension of the gas in the alveoli of the lungs were maintained at unity and that in the blood at zero.

Information of this sort, while of comparatively little physical value, might be of very great physiological usefulness. For a gas such as oxygen, for example, it is usually not so important to know the respective parts played in its intake by such factors separately as the total area, the thickness, etc., of the alveolar walls as how much oxygen would, in fact, be taken up under some given conditions. Information of the latter sort is given by the constant, $k$. Other things being equal, an individual in whom $k$ is large would be expected to fare better at high altitudes, where the effects of oxygen-lack tend to appear than one in whom $k$ is small. That such a relation exists is indicated by the work of Barcroft et. al. (1922). It is true that in the case of oxygen, direct determinations of $k$ by the method described above are usually impracticable because its tension in the blood is not only not zero, but it is not even constant in different parts of the capillary bed where absorption occurs. By an application, however, of the law discovered by EXNER (1875) that the rates of diffusion of two gases through a membrane are directly proportional to their solubilities and inversely proportional to the square roots of their molecular weights, M. KROGH (1915) was able from actual determinations of $k$ for CO to obtain a calculated value of $O_2$; it is this calculated value that is usually employed, though BARCROFT et al. (1920) obtained a rough estimate of such a value by a more direct method.

In recent years the rate of diffusion of solutes from a given quantity of solution through a membrane into pure water has been used in investigating the factors governing the permeability of membranes. FUJITA (1926) in a study of this sort measured the relative amounts of different substances, as compared with a standard substance, urea, that crossed a dried collodion membrane in a given time. This method of comparison of diffusion rates is unsatisfactory, however, except for the purposes of arranging substances in the order of their diffusibility, since the ratios so obtained obviously depend upon the time arbitrarily selected for the measurements; all such ratios must approach unity as the time is indefinitely increased. A much better method of comparison is to apply equation (72) above to the case where the external concentration $c$ is kept approximately equal to zero. The equation for this case becomes

$$u = u_0 e^{-Kt}. \tag{74}$$

This equation with the substitution of $\lambda$ (called the dialysis constant) for $K$ has been extensively used by BRINTZINGER (1927—1932). The important fact appears in BRINTZINGER's work that in many cases the dialysis constant bears the same relation to the molecular weight as does the true diffusion coefficient, $D$. Where this relation holds, the dialysis method has many practical advantages as compared with the method of free diffusion. It must be remembered, however, that a simple relation between $\lambda$ and $D$ can be expected only when no factor is present which has not been taken into account in the derivation of the equation. If, for example, in a porous membrane the pores, or some of them, are sufficiently small to exclude, or to admit with difficulty large molecules, while not hindering to the same extent the diffusion of small ones, then the observed rates of dialysis of different substances might greatly exaggerate the differences shown by the method of free diffusion. This is, in fact, what BRINTZINGER, in agreement with FUJITA, finds to be the case with certain types of membranes and substances.

### $\beta$) Diffusion of water alone.

The diffusion of water through membranes is of enormous importance, both in the physiology of the cell and of the entire organism, but it cannot be treated by means of FICK's law, which relates the rate of diffusion to the concentration gradient of the diffusing substance. It will be remembered that, strictly speaking, this law can be expected to hold accurately only for very dilute solutions, in which the molecules of the diffusing substance are not merely distributed in a discontinuous manner, but in which they are so far separated as to have no influence on each other's movements. In the case of water, however, the concentration in the solutions ordinarily dealt with in physiology is very high — of the order of magnitude of 56 M — and the individual molecules are so close together that they form a continous body, which because of the cohesive forces in the liquid is capable of behaving in certain respects as a single unit. To attempt to treat a case of this sort by laws appropriate to dilute solutions would clearly be unwarranted. It happens, however, that an equation very similar to FICK's, in which osmotic pressure takes the place of concentration, can be applied under these conditions.

Such an equation may very readily be derived for the case that has been studied most thoroughly, namely, that of a watersoaked membrane separating two solutions of a solute to which the membrane is impermeable. Let the thickness of such a membrane be $H$ and assume that it is so rigid that it can undergo no volume changes. Let the osmotic pressure of the more concentrated solution be $p_2$ and that of the more dilute solution $p_1$. In such a system the escaping tendency of the water will be highest in the water-soaked membrane itself, and water will therefore tend to pass from that region into both of the others. But, by hypothesis, the volume of the membrane is fixed, and the cohesive properties of water are such that forces of the magnitudes involved cannot create a vacuum within the system. It follows, therefore, that the force tending to cause a passage of water outward from the membrane into either solution must at the same time tend to bring water inward from the other solution; the observed effect will be the resultant of the two sets of forces. The forces in question can be measured by determining the magnitudes of opposing forces which would just balance them. Since osmotic pressure may be defined as the pressure required to make the escaping tendency of the solvent in a given solution equal to that of the pure solvent under otherwise similar conditions, and since pressure is force per unit area, a knowledge of the osmotic pressures of the two solutions gives a measure of the effective driving force within the system. If the amount of water acted upon by this driving force be equal to the volume of the membrane multiplied by the fraction of this volume, $\varphi$, occupied by water and by the number, $n$, of units of quantity of water contained in unit volume, we have for the force exerted on one unit of quantity, (since the area, $A$, cancels out)

$$k = \frac{p_2 - p_1}{H n \varphi}.$$

The velocity of movement, $v$, of the entire amount of water in the membrane is equal to $k/R$ where $R$ is the resistance against which each unit of quantity moves at unit velocity. The amount that will cross the boundary of the membrane in unit time is therefore equal to $v A \varphi n$ and we have finally

$$\frac{dQ}{dt} = \frac{A}{R} \frac{p_2 - p_1}{H}. \tag{75}$$

This equation resembles one form of FICK's equation given above [(14), p. 31] but, unlike it, is entirely general, and is not limited to very thin membranes or to steady states in thick ones. The reason for the wider applicability of equation (75) is that in the diffusion of water across a membrane the diffusing substance is continuous and behaves as a single unit, instead of consisting of a very large number of completely independent units.

If equation (75) be assumed to be valid for the plasma membranes of living cells, it becomes very easy to study in a quantitative way the permeability of such cells to water. This study is facilitated by the fact that the volume changes of cells in anisotonic solutions may for practical purposes be considered to be equal to the volumes of the water which enter or leave the cells. By merely measuring at intervals the diameters of spherical cells, a means is provided for determining $Q$ in equation (75). Of the other quantities in this equation, $A$, the area of the cell is calculated from the diameter. $H$, the thickness of the membrane, is not known, but may be incorporated as described above (p. 74) with the diffusion coefficient and, if necessary, the partition coefficient, to form a permeability constant for water, $k_2$. The external osmotic pressure, $p_1$, may be kept constant by using an excess of solution. Preferably the external solution should be kept stirred, but even if it is not, the rate of diffusion in it is usually so much faster than that across the cell membrane, that little error results from considering the outside of the membrane to be in contact with a solution of constant composition. Within the cell, diffusion is also usually sufficiently rapid so that the internal concentration may, at least as a first approximation, be considered to be uniform.

Assuming that the change in volume of a cell is equal to the volume of the water which crosses its boundaries, that osmotic pressure is directly proportional to concentration, that mixing inside and outside the cell membrane is complete and continuous, that the entire cell volume takes part in the osmotic exchange, and that the VAN'T HOFF law may be applied to the system, equation (75) may be written

$$\frac{dV}{dt} = k A (p - P) \tag{76}$$

where $p$ and $P$ are the internal and external osmotic pressures, respectively. In one of the first attempts made to represent the volume changes of cells in anisotonic solutions by a mathematical law, LILLIE (1916) used essentially this equation. By an error in its integration, however, he was led to an equation which, though it seemed to give a fairly good representation of his own observations and of the earliest of the extensive and important series made by LUCKÉ and McCUTCHEON (1932), was nevertheless misleading in that the value of $k$ obtained by it was not the same for different external concentrations. This mistake was rectified by NORTHROP (1927) who gave the correct integrated form of the equation which has been used in all subsequent work. About the same time, JACOBS (1927) independently used the following similar equation for the swelling of an erythrocyte in a hypotonic solution, taking advantage of the practical constancy of the surface of this kind of cell during the early stages of the swelling process

$$k = \frac{p_0 V_0}{P^2 A t} \ln \frac{p_0 V_0 - P V_0}{p_0 V_0 - P V} - \frac{V - V_0}{P A t}. \tag{77}$$

To this equation which closely resembles the one obtained by NORTHROP may be added the corresponding one for swelling in water

$$k = \frac{V^2 - V_0^2}{2 p_0 V_0 A t}. \tag{78}$$

By means of these equations, JACOBS determined a value of $k$ for the human erythrocyte that indicates the passage of about 3 cubic micra of water per square micron of surface per minute per atmosphere of difference in osmotic pressure (see also JACOBS 1932). To this value may be added one about 30 times as small for the Arbacia egg obtained by LUCKÉ, HARTLINE and McCUTCHEON (1931) one approximately 120 times as great for the wall of the capillary of the frog's mesentery obtained by LANDIS (1927), and a series of values somewhat greater than that for the Arbacia egg reported by LEITCH (1931) for other echinoderm eggs. In an extensive and accurate series of measurements HÖFLER (1930) and HUBER and HÖFLER (1930) have applied an equation very similar to (77) to large numbers of plant cells, but because of the fact that they failed to take into account the cell surfaces and the absolute volumes of their cells, their results, for the present, cannot be compared with those of other workers. For a full discussion of this important subject the review article by LUCKÉ and McCUTCHEON (1932) should be consulted [see also MANEGOLD and STÜBER (1933)].

### γ) Diffusion of a solute and water together.

The cases involving the penetration of a membrane by a solute alone and by water alone having been dealt with, that in which both processes occur simultaneously may next be considered. In living cells it frequently, and indeed under certain conditions necessarily, happens that a movement of water across the cell membrane accompanies that of a solute. The cell is a delicately balanced osmotic system, at all times permeable to water, and any increase or decrease in its internal osmolar concentration must inevitably bring about volume changes except to the extent that swelling is prevented as for example in plant cells by an external cellulose cell wall.

One practically important consequence of the close association of osmotic movements of water with the diffusion of solutes in living cells is that the former, which can frequently be observed and measured with ease, may be used as a means of investigating the latter, which are usually invisible and therefore difficult to follow continuously. The method of osmotic volume changes has indeed for many years been one of the most important methods for studying the permeability of living cells. Following the early observation of DE VRIES (1871) and KLEBS (1888) on the recovery in the presence of penetrating substances of the original volume by plant cells after plasmolysis, a study of the rate of deplasmolysis became one of the favorite methods for the estimation of the permeability of cells to different solutes. For references to the extensive literature on this subject HÖBER (1926), STILES (1924), GELLHORN (1929), JACOBS (1924), BÄRLUND (1929), etc. may be consulted.

The method in question as commonly employed, however, has several defects. The first is that while, in general, it permits a comparison of the penetration of the same cell by different substances, enabling the latter to be arranged in the order of their rates of penetration, it is useless without further refinements for making comparisons between different cells having different surfaces and volumes, since the time required for a given exchange by diffusion or osmosis, or both, obviously depends to a high degree upon the dimensions of the system involved. The second defect of the older methods is that whereas in the case of slowly penetrating solutes the rate of penetration of water in comparison with that of the solute may without great error be assumed to be infinite, thereby implying at all times an osmotic balance between the cell and its surroundings, this assumption is by no means justified in the case of more rapidly penetrating substances where osmotic balance is not necessarily maintained, and where the volume of the cell therefore bears no simple relation to the amount of solute that has entered it.

Several attempts have been made to remove one or both of these objections and to make the method of osmotic volume changes suitable for the quantitative measurement of the permeability of a cell to a solute. Such attempts, unlike those mentioned above, in which the volume of the cell is assumed to remain constant, take advantage in one way or the other of the measured volume changes that occur in solutions of a penetrating substance. Thus, LEPESCHKIN (1908) measured at two successive times the volumes of a cell undergoing deplasmolysis in a solution of a penetrating substance, glycerol. Knowing the concentration of the external solution, and assuming a condition of constant osmotic balance, which implies a very rapid penetration of water in comparison with the solute, he calculated the amount of solute that must have entered the cell to bring about the observed volume change. He then divided this amount by the average area of the cell and the average concentration gradient across its membrane during the period of the experiment and in this way obtained a permeability constant for the solute in question.

The method of LEPESCHKIN, while applicable to cases in which the movement of water is rapid in comparison with that of the solute, must necessarily fail when this relation no longer holds, as frequently happens in the case of animal cells. Thus, STEWART (1931) wished to determine whether the increase of permeability to water of the Arbacia egg on fertilization (LILLIE 1916) is paralleled by a corresponding increase in permeability to a harmless and rapidly penetrating solute, ethylene glycol. Experiments designed for the purpose showed that the rate of swelling in solutions of this substance isosmotic with sea water was indeed greater after fertilization than before, but by the methods then available it was impossible to determine satisfactorily how much of the difference was due to an increased permeability to the solute and how much to the change already known to occur in the case of water. In an attempt to deal more readily with cases of this sort JACOBS and STEWART (1932) employed the following very simple method.

Spherical cells, such as the eggs of Arbacia, whose changes in diameter can be measured continuously be means of an ocular micrometer are placed in their natural medium whose known osmolar concentration is $C_M$ and to which a penetrating solute has been added in the osmolar concentration $C_S$. The resulting hypertonic solution at first causes a shrinkage of the cells, followed by a recovery of their initial volumes. An important advantage of using an external medium of this sort is that it contains all the electrolytes to which the cell is accustomed in exactly their normal concentrations, and permeability is therefore studied under conditions as nearly natural as possible. A second, mathematical rather than physiological, advantage is that the observed time-volume curve passes through a minimum where $\frac{dV}{dt} = 0$; at this point the differential equation describing penetration undergoes a very useful simplification.

It may be assumed that the assumptions already made are applicable to a system of this sort and in addition that the outward flow of water across the cell membrane does not significantly prevent the inward passage of the solute. This latter point will require further investigation; but the existing evidence, as far as it goes, seems to justify such an assumption, at least as a working hypothesis. Under these conditions equations (71) and (76) may be written, using the symbols employed by JACOBS and STEWART, in the following form

$$\frac{dS}{dt} = k_1 A \left(C_S - \frac{S}{V}\right) \tag{79a}$$

$$\frac{dV}{dt} = k_2 A \left(\frac{a+S}{V} - C_S - C_M\right) \tag{79b}$$

where, in addition to the symbols already explained, $S$ represents the amount of solute that has entered the cell up to time $t$, $a$ the amount of osmotically active, non-diffusible materials originally present in the cell, $V$ the cell volume and $k_1$ and $k_2$ permeability constants for the solute and water, respectively. It may be assumed that changes in cell volume are directly proportional to the amounts of water that have crossed the cell membrane and that osmotic pressure is directly proportional to concentration, within the range employed. Both of these assumptions are sufficiently accurate for practical purposes. It will be noted that the numerical values of the two constants, $k_1$ and $k_2$, depend upon the units of volume, area and concentration employed. Any convenient units may be used, but in making comparisons between different cells it is necessary to be certain that the units are the same in each case.

Equation (79b) may now be rearranged as follows

$$S = \frac{V}{k_2 A} \frac{dV}{dt} + (C_M + C_S) V - a.$$

In general, the relation between $S$ and $V$ is a complicated one, but for one particular volume, namely the minimum volume, the relation is extremely simple, i. e.,

$$S = (C_M + C_S) V_{\text{min.}} - V_0 C_M \tag{80}$$

where instead of $a$ is written the product of the initial volume and the osmolar concentration of the medium with which it is in equilibrium. By means of equation (80) it is easy to determine that in a certain time (that required for the attainment of the minimum volume) a known amount of solute has entered the cell. Presumably its entrance has been governed by equation (79a). If $V$ and $A$ in this equation were constant, the value of $S$ as found above could be introduced into it and the equation could be integrated, giving, after evaluation of the integration constant

$$k_1 = \frac{V}{At} \ln \frac{C_S V}{C_S V - S}. \tag{81}$$

Though $V$ and $A$ are not constant, and strictly speaking this integration is unjustified, it is at any rate certain that the true value of $k_1$ must lie somewhere between those obtained by taking successively $V = V_0$ and $V = V_{\text{min.}}$ using at the same time the corresponding values of $A$ in equation (81). It is easy to show that for the volume changes actually involved in the studies on Arbacia eggs the difference between these extreme values rarely exceeds 10 per cent. The distance of the true value of $k_1$ from either extreme would be less than this, and the distance from a value calculated from the arithmetical mean of both would be still less. Since differences of this magnitude are negligible in systems where unavoidable errors of measurement are even greater, it is possible, therefore, to employ equation (81) as it stands.

By the use of this equation, STEWART and JACOBS (1932a) were able to show that increase in permeability of the Arbacia egg to ethylene glycol on fertilization, which was indicated but not certainly proved by the earlier method of analysis, not only occurs but is of the same order of magnitude as the increase in permeability to water. They also showed (STEWART and JACOBS 1932b) that while an increase of temperature increases the permeability of the unfertilized egg to both water and to ethylene glycol, the temperature coefficient is much higher for the latter substance than for the former, and also (STEWART and JACOBS 1933) that the penetration of the two substances is very differently affected by changes in the ionic constitution of the surrounding medium. It is of interest that the permeability constants obtained in this way by JACOBS and STEWART for several different solutes with the Arbacia egg agree very closely with those obtained by a direct chemical method for an entirely different type of cell *(Chara)* by COLLANDER and BÄRLUND (1933). It will be noted that the method just described, though involving at one point in the mathematical treatment the not entirely accurate assumption of a constant volume and surface, in this respect resembling several of those employed by previous workers, has the important advantage over the latter that it can be applied to rapidly penetrating solutes. It was indeed for use with such solutes that it was developed.

For more accurate work and for theoretical purposes in general, it would be advantageous if the inexact assumption of constant volume and surface could be omitted and if the simultaneous differential equations (79 a) and (79b) describing the behavior of the cell and of the penetrating solute could be solved as they stand. This is possible by various approximate numerical methods, and such a solution was obtained by JACOBS (1933c) for the particular case of an Arbacia egg exposed to an 0.5 M solution of a penetrating non-electrolyte in sea water, as well as for other cases in which $C_M$ and $C_S$ have the same ratio as in the one mentioned. The solution obtained takes the form of a chart from which may be read off, not only the value of $k_1$ corresponding to the time of attainment of some particular minimum volume, but the value of $k_2$, the permeability of the same cell to water as well. The relative parts played by the two constants in determining the magnitude and the time of attainment of the minimum value is discussed in another paper by JACOBS (1933b). It is there shown that while the time at which the minimum volume is reached depends on the absolute magnitudes of the two constants, the value itself is determined only by their ratio $K = k_2/k_1$. For the most important case where $C_M$ is taken as equal to unity but where $C_S$ may have any value, the theoretical minimum volume may be found by means of the equation

$$\ln (V_{\min} - 1)^2 = \ln K C_S^2 + \frac{K C_S + K + 1}{\sqrt{A}} \ln \frac{K C_S + K + 1 - \sqrt{A}}{K C_S + K + 1 + \sqrt{A}}$$

where $$A = 4 K C_S + (K C_S + K - 1)^2. \tag{82}$$

Other equations are also derived which show the relation between the volume of the cell and the amount of solute that has penetrated it, etc. For further details the original papers may be consulted.

A somewhat similar problem has recently been treated by LONGSWORTH by a different method, which is preferable in that it permits the final differential equations that describe the process to be integrated directly and that the treatment, in general, follows strictly thermodynamic lines. The present method, however, has the advantage of greater simplicity in the manner of formulation of the necessary differential equations, and shows more clearly its relation to other published work on cell permeability.

If in equations (79a) and (79b) the area of the cell could be considered to remain constant while its volume changed, certain simplifications of the mathematical treatment would be possible. Such a condition actually exists in the mammalian erythrocyte during the early stages of a swelling process, since the cell has a biconcave form and the two concavities may be converted into convexities with no change in surface. Taking advantage of this fact, JACOBS (1934) has derived equations and calculated tables which permit the estimation of the rates of entrance of rapidly penetrating solutes into the mammalian erythrocyte from measurements of the time of attainment of the hemolytic volume in water, on the one hand, and in an isosmotic solution of the penetrating solute, on the other. It is shown in the same paper that when the time of osmotic hemolysis in an approximately 0.3 M solution of a non-electrolyte exceeds a few minutes, no appreciable error is caused by considering the rate of entrance of water to be infinitely great and using a simpler method of treatment. Such a method, involving the assumption of an infinite rate of penetration of water into the erythrocyte, has recently been employed by SCHIØDT (1933). His equations at first sight seem to differ from those derived from equations (79a) and (79b) in that they contain a factor for the initial cell volume. This difference, however, is only an apparent one, since the factor in question in the equations of JACOBS was for convenience, by the use of different units, incorporated in the constant $k_1$. This constant may at will be converted into one involving any other desired units, and such a change will, in general, be made for purposes of comparison between different cells (see in this connection JACOBS 1934).

## 10. One-dimensional diffusion processes in infinite and semi-infinite systems.

By an infinite system, in cases of diffusion in one dimension, will be understood a system that has no boundaries which can be reached or crossed by the diffusion stream; by a semi-infinite system one that has a single boundary of this sort. In the case of the homogeneous finite systems so far considered there have always been two such boundaries. Strictly speaking, infinite and semi-infinite systems are merely a convenient mathematical fiction, never encountered in actual experience; but, because of the slowness of diffusion processes, whenever the distance involved is great or the time is short, or both, real systems may for practical purposes be treated as if they were infinite or semi-infinite in extent. Examples of such treatment are not only very common in connection with the measurement of diffusion coefficients by physicists but will be found in the recent physiological literature as well.

When it is possible to treat an actual system as if it were infinite, certain practical and theoretical advantages result. In the first place, mathematical computations are much simplified; instead of requiring the somewhat laborious evaluation of a series of sine-exponential or cosine-exponential terms, the solution of a problem involving an infinite or semi-infinite system may frequently be obtained directly from a table of values of the normal probability integral—indeed, some of the most important problems connected with such systems give rise to equations which can be solved by purely arithmetical means [see, for example, numbers (103) and (104) below]. So great are the mathematical advantages of the methods about to be described that STEFAN (1879) and others have preferred to apply them, with the assistance of the principle of reflected diffusion streams (p. 18), even to systems in which very considerable amounts of the diffusing substance reach the boundary or boundaries, and to which FOURIER's method of analysis is entirely applicable.

A theoretical advantage of the diffusion equations for infinite systems is that they are, in general, very closely related to those describing the random Brownian movement of suspended particles or molecules in solution, which will be discussed in the following section. They therefore correspond to something that has a real existence, of which they give a readily intelligible mathematical picture. A FOURIER's series, on the other hand, though it may have a sufficiently concrete meaning in the case of vibrating strings, possesses only a highly artificial relation to a diffusion process. Other things being equal, therefore, the methods about to be described will usually be employed in cases where they provide a possible alternative to those considered in the preceding sections.

**a) Solution of the diffusion equation for infinite systems.**

In all discussions of FOURIER's series (see, for example, BYERLY 1893, p. 52) it is shown, by the application of very simple trigonometric principles, that as the range, $H$, within which solution applies, is increased without limit, the general FOURIER's series passes over into a definite integral known as FOURIER's integral

$$f(x) = \frac{1}{\pi} \int_{-\infty}^{\infty} f(\lambda)\, d\lambda \int_{0}^{\infty} \cos \alpha (\lambda - x)\, d\alpha \tag{83a}$$

in which $\lambda$ and $\alpha$ are merely variables of integration which disappear when for them are substituted the values of the limits of the two definite integrals. Equation (83a) may also be written in the alternative form

$$f(x) = \frac{1}{\pi} \int_{0}^{\infty} d\alpha \int_{-\infty}^{\infty} f(\lambda) \cos \alpha (\lambda - x)\, d\lambda . \tag{83b}$$

When $f(x)$ is an odd function of $x$; that is, when $f(x) = -f(-x)$ then (83 a) may readily be transformed (see BYERLY l. c., p. 54) to

$$f(x) = \frac{2}{\pi} \int_{0}^{\infty} f(\lambda)\, d\lambda \int_{0}^{\infty} \sin \alpha \lambda \sin \alpha x\, d\alpha . \tag{84a}$$

Similarly, when $f(x)$ is an even function of $x$; that is, when $f(x) = f(-x)$

$$f(x) = \frac{2}{\pi} \int_{0}^{\infty} f(\lambda)\, d\lambda \int_{0}^{\infty} \cos \alpha \lambda \cos \alpha x\, d\alpha . \tag{84b}$$

Although (84a) holds for all values of $x$ when $f(x)$ is odd and (84 b) holds for all values of $x$ when $f(x)$ is even, either equation may be used when only positive values of $x$ are involved, just as in the case of the original FOURIER's series (see p. 35 above).

In dealing with diffusion in one dimension in an infinite system it is necessary, as before, to find a solution of FICK's equation which for all values of $x$ and $t$ is compatible with the conditions of some particular problem. Since in an infinite system there are no boundary conditions to be considered, it is sufficient that the solution should satisfy merely the initial condition

$$u = f(x) \quad \text{when} \quad t = 0 .$$

As before, particular solutions of the general diffusion equation must be selected and then properly combined; in the case of an infinite system it is necessary that for $t = 0$ the sum of these solutions shall reduce to one of FOURIER's integrals which represents $f(x)$ over the range $-\infty$ to $+\infty$. As possible particular solutions we may take, as previously,

$$u = \sin \alpha x\, e^{-\alpha^2 D t} \quad \text{and} \quad u = \cos \alpha x\, e^{-\alpha^2 D t}.$$

These are still solutions when multiplied respectively by $\sin\alpha\lambda$ and $\cos\alpha\lambda$, since the latter expressions involve neither of the variables $x$ or $t$. Adding $\sin\alpha\lambda\sin\alpha x\, e^{-\alpha^2 Dt}$ and $\cos\alpha\lambda\cos\alpha x\, e^{-\alpha^2 Dt}$ and employing the well-known trigonometric formula

$$\sin x \sin y + \cos x \cos y = \cos(x - y)$$

we have

$$u = e^{-\alpha^2 Dt} \cos\alpha(\lambda - x)$$

which must still be a solution of FICK's equation and will continue to be when multiplied by $f(\lambda)\,d\lambda$. Furthermore, since integration is merely the limit of a process of summation of a series of terms of the same form

$$u = \int_{-\infty}^{\infty} e^{-\alpha^2 Dt} f(\lambda) \cos\alpha(\lambda - x)\,d\lambda$$

is also a solution. We may finally multiply this integral by $\frac{1}{\pi}$ and by $d\alpha$ and again by integration with respect to $\alpha$ take the limit of a series of terms having different values of $\alpha$ but the same form, obtaining

$$u = \frac{1}{\pi}\int_{0}^{\infty} d\alpha \int_{-\infty}^{\infty} e^{-\alpha^2 Dt} f(\lambda) \cos\alpha(\lambda - x)\,d\lambda. \tag{85}$$

This equation is still a solution of FICK's equation and when $t = 0$ it reduces to FOURIER's integral (83 b) which represents $f(x)$ over the range $-\infty$ to $+\infty$. It is therefore the desired solution of our problem.

Equation (85) may be made more useful for present purposes, first, by taking its alternative form analogous to (83 a), and then by applying to the latter the standard form for integration (see any table of definite integrals)

$$\int_{0}^{\infty} e^{-a^2 x^2} \cos bx\,dx = \frac{\sqrt{\pi}}{2a} e^{-\frac{b^2}{4a^2}}$$

giving finally as the fundamental equation for diffusion in infinite systems

$$u = \frac{1}{2\sqrt{\pi D t}} \int_{-\infty}^{\infty} f(\lambda)\, e^{-\frac{(\lambda - x)^2}{4Dt}}\,d\lambda. \tag{86}$$

### b) Diffusion from a very thin layer.

Equation (86) may now be applied to several initial types of distribution of material represented in the equation by the form of $f(\lambda)$. The simplest possible case is that of an extremely thin layer of solution of concentration $u_0$ and of thickness $\delta x$ initially placed in an infinite column of solvent at the level $x = 0$. Assuming a cross-section of unity, the total amount of diffusible material in the system is $u_0\,\delta x$.

Though $\delta x$ is very small, $u_0$ may, in theory, be made sufficiently great to account for any desired initial amount of substance. In general, for an initial condition $u = f(x)$ when $t = 0$, the total amount of diffusible substance is $\int_{-\infty}^{\infty} f(x)\, dx = \int_{-\infty}^{\infty} f(\lambda)\, d\lambda$; mathematically it may be thought of as distributed in an infinite number of layers, each of thickness $dx$, and having a concentration whose relation to the position of the layer is indicated by the form of $f(\lambda)$. In the present case within a narrow region whose position does not differ appreciably from $\lambda = 0$ we have the constant concentration $u_0$; elsewhere the concentration is zero. Equation (86) for this particular case therefore reduces to the form

$$u = \frac{u_0\, \delta x}{2\sqrt{\pi D t}} e^{-\frac{x^2}{4Dt}}. \tag{87}$$

That is to say, at any time, $t$, the concentration at any distance, $x$, from the original thin layer of diffusible material will depend in the manner indicated by equation (87) upon the amount of diffusible material $u_0\, \delta x$, which may for convenience be represented by $q$. Furthermore, the amount contained at any time $t$ in a layer of unit cross-section lying between $x$ and $x + dx$ will be $u\,dx$; and the total amount lying between $x = 0$ and $x = a$ must therefore be the sum of all such infinitesimal layers between the limits in question, namely,

$$Q_{0,a} = \frac{q}{2\sqrt{\pi D t}} \int_0^a e^{-\frac{x^2}{4Dt}}\, dx. \tag{88}$$

This equation for practical purposes may be improved by introducing a new variable $z = \frac{x}{2\sqrt{Dt}}$ which causes it to assume the form

$$Q_{0,a} = \frac{q}{\sqrt{\pi}} \int_0^{\frac{a}{2\sqrt{Dt}}} e^{-z^2}\, dz. \tag{89}$$

For greater compactness a symbol $\psi(x)$ may be used to represent the value of the normal probability integral for the argument $x$. Remembering that the diffusion process is symmetrical, or alternatively, after introducing into equation (88) a lower limit $-a$, and remembering that the integral itself is symmetrical about $x = 0$, we obtain the useful relation

$$Q_{\pm a} = q\,\psi\left(\frac{a}{2\sqrt{Dt}}\right). \tag{90}$$

By way of illustration, equation (90) may be applied to the following problem which will serve to illustrate in a concrete way certain important aspects of the diffusion process. Suppose that 1,000,000 molecules of cane sugar, could by some means be placed at the bottom of tall cylindrical vessel containing water. What would be their distribution at the end of an hour? The solution of the problem may conveniently take the form of a calculation of the numbers of molecules that would be found in successive layers of 1 mm. in thickness. Though equation (90) gives the theoretical number of molecules lying between the limits $\pm a$, it is nevertheless applicable without change to the present one-sided distribution, since all the molecules that, if free to do so, would pass in the negative direction must here be reflected upward by the bottom of the vessel. The value of $D$ for cane sugar may be taken as 0.33 cm.$^2$/day; $t$ must therefore be expressed in fractions of a day. For the desired time, equation (90) is successively applied to values of $a$ of 0.1, 0.2, 0.3, etc., using for this purpose an ordinary table of the probability integral. By the subtraction of each value of $Q$ from the one that follows it, approximate figures such as those in the table are obtained:

| Distance from the bottom of the vessel mm. | Number of molecules |
|---|---|
| 0—1 | 453,500 |
| 1—2 | 318,700 |
| 2—3 | 157,350 |
| 3—4 | 54,590 |
| 4—5 | 13,290 |
| 5—6 | 2,280 |
| 6—7 | 170 |
| over 7 | 20 |

This method of treatment may also be applied to finite systems. For example, suppose that diffusion in the case just discussed had been allowed to take place, not in a very tall column of liquid, but in one only 4 mm. high. The distribution in such a case at the end of 1 hour could readily be obtained from the figures just given by applying the reflection principle described on p. 18. It is easily seen that for layers 1 mm. in thickness the number of molecules in each layer from the bottom upward would be 453,520; 318,870; 159,630; and 67,880.

Cases of diffusion from a very thin layer are not limited to those of a purely imaginary nature. For example, a very important study was made by WESTGREN (1914) in which the principles just discussed enabled him to determine $D$ for colloidal gold and selenium preparations and from $D$, in turn, to deduce other properties of the colloidal particles. The method used was to throw down the particles by means of centrifugal force into a layer on the bottom of a chamber in which, as they diffused upward, they could be counted at any desired level by the use of the ultramicroscope. By equation (87) it is obvious that in such a system the concentrations of the particles, i. e. their numbers in a given volume at any two levels, $x_1$ and $x_2$, must bear to each other the relation

$$\frac{u_2}{u_1} = e^{-\frac{x_2^2 - x_1^2}{4\,Dt}}.$$

Taking the logarithms of both sides of the equation, it is evident that if $\ln u$ be plotted against $x^2$, a straight line ought to result. That such straight lines are in fact obtained and may be used for the estimation of $D$ will be seen by consulting the various graphs given by WESTGREN, which are based upon such figures as the following typical ones, selected at random from his published data.

| Distance in $\mu$ from the bottom of the vessel | Concentration of particles | $Log_{10}$ of Concentration |
|---|---|---|
| 50 | 595 | 2.775 |
| 250 | 568 | 2.755 |
| 450 | 476 | 2.677 |
| 650 | 371 | 2.569 |
| 850 | 271 | 2.433 |
| 1050 | 165 | 2.217 |
| 1250 | 90 | 1.956 |
| 1450 | 45 | 1.649 |

Another application of the behavior of a very thin layer of solution in a tall column of water is found in the ingenious method suggested by PROCOPIU (1918) for determining the diffusion coefficients of electrolytes. The principle of this method is to measure continuously the potential difference between two electrodes, one of which is placed so far from an original thin layer of solution of a diffusing substance at the bottom of the diffusion vessel that practically no diffusing material reaches it, and the other at any convenient position much nearer to the bottom of the vessel. Theory demands, and experience shows, that the potential difference between the electrodes in such a system must at first increase and then, after passing a maximum, decrease as the concentration at the second electrode passes through the same phases. That the concentration at any given finite level above the plane of contact between the thin layer of solution and the solvent must, in fact, pass through a maximum is obvious from the circumstance that it starts at zero and again tends to return to zero as the diffusing material distributes itself more and more widely in an infinite quantity of liquid. The time at which the maximum is reached at any given level is readily found by a simple application of the principle of maxima and minima to equation (87). Evidently $u$ must be a maximum at the finite time for which $\frac{\partial u}{\partial t} = 0$. Performing this differentiation, equating the resulting expression to zero and solving for $t$ we obtain

$$t_{\text{max.}} = \frac{x^2}{2D}. \tag{91}$$

From this very simple relation $D$ can be obtained from accurate measurements of time and distance alone. Absolute values of potential difference and concentration do not enter into equation (91); it is only necessary to know when at a given level they reach their maximum values. It will be noted that in this equation the usual relation between time and the square of a distance of diffusion appears.

### c) Diffusion from a layer of finite thickness.

The case just discussed assumes several forms, but the most important of these practically is that of diffusion from a very thin layer originally placed at the bottom of an extremely (in theory, infinitely) tall column of liquid. The related case may now be considered of the behavior under similar conditions of an initial layer of solution of greater but still of finite thickness, $h$. This problem may be solved in several different ways. The one that will first be discussed depends upon the fact that in the region from $x = 0$ to $x = \infty$ it makes no difference whether the system be semi-infinite with a closed boundary at $x = 0$, and a layer of solution occupying the space between $x = 0$ and $x = h$, or infinite with a layer of solution of thickness $2h$ occupying the space between $x = -h$ and $x = +h$. Since the infinite system is symmetrical, there can occur in it no detectable passage of material across the plane $x = 0$, which will therefore behave exactly as if it were a closed partition. In a system of this latter type in which the initial condition is $u = u_0$ between $-h$ and $+h$ and $u = 0$ elsewhere, we have by equation (86)

$$u = \frac{u_0}{2\sqrt{\pi D t}} \int_{-h}^{h} e^{-\frac{(\lambda - x)^2}{4 D t}} d\lambda .$$

A slight change in the variable enables this integral to be converted into a form similar to that found in tables of the normal probability integral. Let $\beta = \frac{\lambda - x}{2\sqrt{Dt}}$; then $\lambda = x + 2\sqrt{Dt} \cdot \beta$ and $d\lambda = 2\sqrt{Dt}\, d\beta$; also, the previous upper and lower limits of integration now become $\frac{h - x}{2\sqrt{Dt}}$ and $-\frac{h + x}{2\sqrt{Dt}}$. Making these changes we have

$$u = \frac{u_0}{\sqrt{\pi}} \int_{-\frac{h+x}{2\sqrt{Dt}}}^{\frac{h-x}{2\sqrt{Dt}}} e^{-\beta^2} d\beta . \tag{92}$$

Finally, remembering that the normal probability integral is symmetrical on the two sides of $x = 0$ the equation becomes

$$u = \frac{u_0}{2}\left[\psi\left(\frac{h + x}{2\sqrt{Dt}}\right) + \psi\left(\frac{h - x}{2\sqrt{Dt}}\right)\right]. \tag{93}$$

The value of $u$ corresponding to any values of $x$, $t$ and $D$ may now be found with great ease.

A very similar equation, obtained by another method, was used by STEFAN (1879) to calculate the amounts of material that in any given time, for different values of $D$, would pass from an initial layer of solution of concentration $u_0$ and height $h$ into a series of layers of any desired thickness in an infinite overlying column of liquid. It is, of course, evident that equation (93) gives concentrations rather than amounts, but the latter are readily obtained from the former. If the layers in question be sufficiently thin, little error results from considering that the average concentration within each layer is that at its mid-plane and multiplying this concentration by the volume of the layer. For greater accuracy, especially if the layers be thick, it is necessary to calculate enough values of $u$ to evaluate the integral $\int_{x_1}^{x_2} u\,dx$ by means of SIMPSON's rule or some analogous method. The necessary labor, however, even in this case is not great. In fact, so satisfactory is this method that STEFAN applied it even to finite systems to which the FOURIER method of analysis is also applicable. As has already been mentioned, in order to apply the calculated figures for an infinite system to a finite system it is only necessary to use the principle of reflection discussed above on p. 18.

### d) Diffusion from a column of unlimited length.

A third, and most important, type of infinite system is that in which when $t = 0$, $u = u_0$ from $x = -\infty$ to $x = 0$ and $u = 0$ from $x = 0$ to $x = \infty$. This arrangement may be assumed to exist in finite systems consisting of a layer of solution and of solvent, respectively, each sufficiently thick so that diffusion does not appreciably affect the regions farthest removed from the original plane of contact. Examples of the approximate treatment of finite systems by means of equations derived for this type of infinite system are given by STEFAN (1878a), SCHUHMEISTER (1879), GRAHAM (1904, 1907), CARLSON (1911a, b), BRILLOUIN (1912) (see criticism by FÜRTH 1927c, p. 210), DUMMER (1919), SVEDBERG (1925) and others. In particular, the important micro-method for studying diffusion processes introduced by FÜRTH (1925) employs the same principle. This method has recently been used, among others, by FÜRTH (1925, 1927), FÜRTH and ULLMANN (1927), ULLMANN (1927), NISTLER (1929, 1930, 1931), OSTWALD and QUAST (1929), SÜLLMANN (1931), FISCHER (1931), and ZUBER (1932a, b).

The mathematical treatment of such cases is simple. For them, equation (86) assumes the form

$$u = \frac{u_0}{2\sqrt{\pi D t}} \int_{-\infty}^{0} e^{-\frac{(\lambda - x)^2}{4 D t}}\, d\lambda .$$

Introducing as before a new variable $\beta = \frac{\lambda - x}{2\sqrt{D t}}$ the equation becomes

$$u = \frac{u_0}{\sqrt{\pi}} \int_{-\infty}^{-\frac{x}{2\sqrt{D t}}} e^{-\beta^2}\, d\beta .$$

Remembering the symmetry of the normal probability integral for positive and negative values of $x$, this may be written

$$u = \frac{u_0}{\sqrt{\pi}} \int_0^{\infty} e^{-\beta^2} d\beta - \frac{u_0}{\sqrt{\pi}} \int_0^{\frac{x}{2\sqrt{Dt}}} e^{-\beta^2} d\beta$$

which may be further simplified by remembering that

$$\int_0^{\infty} e^{-\beta^2} d\beta = \frac{\sqrt{\pi}}{2}.$$

The final equation, written in its most convenient form, is

$$u = \frac{u_0}{2}\left( 1 - \frac{2}{\sqrt{\pi}} \int_0^{\frac{x}{2\sqrt{Dt}}} e^{-\beta^2} d\beta \right) \tag{94}$$

or, otherwise expressed

$$u = \frac{u_0}{2}\left[1 - \psi\left(\frac{x}{2\sqrt{Dt}}\right)\right]. \tag{95a}$$

For negative values of $x$, i.e., for positions on the solution side of the original line of separation between solution and solvent, it is very easy by remembering the symmetrical form of the normal probability curve to convert equation (95 a) into

$$u = \frac{u_0}{2}\left[1 + \psi\left(\frac{x}{2\sqrt{Dt}}\right)\right]. \tag{95b}$$

It will be noted that when $t = 0$, equations (95 a) and (95 b) indicate concentrations in the two halves of the system 0 and $\frac{u_0}{2}$ respectively, also that as $t$ approaches $\infty$ both concentrations approach $\frac{u_0}{2}$, and finally that when $x = 0$, $u = \frac{u_0}{2}$ for all values of $t$ greater than zero. An important application of this last property of the system has been made by SVEDBERG (1925) and is referred to below (p. 131).

An important application of equation (95 a) is involved in the method of FÜRTH mentioned above. In this method, following a suggestion of AUERBACH (1924), what is observed is not the concentration existing at a previously determined position at a given time, but rather the position reached by a previously determined concentration at a given time. The advantages of the latter procedure are, first, that one or a few appropriate standards of known concentration of a colored diffusing substance may be prepared in advance, and errors in the indirect calculation of other concentrations are thereby avoided; and, second, that the process of diffusion itself can be studied at any desired concentration with no necessity for assumptions as to the validity of FICK's law.

In cases where equation (95 a) may be considered to be valid, $D$ can readily be obtained by FÜRTH's method, as follows. Let $v$ represent the factor of dilution of the original concentration required to produce the standard concentration whose progress it is desired to follow. For this particular concentration we have

$$\frac{u_0}{v} = \frac{u_0}{2}\left[1 - \psi\left(\frac{x}{2\sqrt{Dt}}\right)\right]$$

$$\psi\left(\frac{x}{2\sqrt{Dt}}\right) = 1 - \frac{2}{v}$$

$$\frac{x}{2\sqrt{Dt}} = \psi^{-1}\left(1 - \frac{2}{v}\right)$$

$$D = \frac{x^2}{t}\cdot\frac{1}{4}\left[\psi^{-1}\left(1 - \frac{2}{v}\right)\right]^{-2} = \frac{x^2}{t} f(v). \qquad (96)$$

A table of values of $f(v)$ has been prepared by Fürth (1927d) and reproduced by Nistler (1931); by means of it the calculation of $D$ is extremely easy. It is obvious from equation (96) that if in the graphic representation of the data obtained from experiments of this type, $x^2$ be plotted against $t$, or $x$ against $\sqrt{t}$, a straight line will result whose slope for each concentration will be different. That this is usually, though not always the case, is apparent from the figures published by Nistler (1929, 1930, 1931), Süllmann (1931), Zuber (1932) etc.

### e) Diffusion in semi-infinite systems.

Semi-infinite systems are of much importance, not only in the measurement of diffusion coefficients and in other physical studies (v. Wroblewski 1877, 1881, Wüstner 1915, Mann 1924, Tammann and Jessen 1929, Dirken and Mook 1930, Bruins 1931, etc.), but in connection with certain physiological problems as well. An example of the latter type of application, which will be discussed more fully below (p. 100), is the case where a muscle or mass of tissue, so large that during the time of the experiment its innermost regions are scarcely affected by diffusion, is exposed to a stirred external medium which preserves at its surface a constant concentration, either higher or lower than that already present in the tissue. Strictly speaking, a case of this sort is likely to involve diffusion in more than one dimension, but when the depth of penetration or of escape of the diffusing substance is small in comparison with the dimensions of the tissue in question, little error is caused in treating the entire surface as if it were a single plane. Problems of this sort are dealt with as if the mass of tissue extended from $x = 0$ to $x = \infty$ and were subject to some specified boundary condition at the plane of contact with the external medium.

The mathematical treatment of semi-infinite systems is, in general, similar to that of the infinite systems discussed above, though it differs in several details. As before, since an infinite distance on one side of the origin is involved, a Fourier integral will be employed, which must reduce to $f(x)$ when $t = 0$. But in addition it must also reduce to zero for $x = 0$. Since in this case, in the system itself, we have to do only with positive values of $x$, any of the integrals (83a), (83b), (84a) or (84b) in the absence of the boundary condition might be employed to represent $f(x)$; however, in order to make the solution reduce to zero when $x = 0$, (84a), which contains $\sin \alpha x$ is chosen. We have, therefore,

$$u = \frac{2}{\pi}\int_0^\infty d\alpha \int_0^\infty e^{-\alpha^2 D t} f(\lambda) \sin \alpha x \sin \alpha \lambda \, d\lambda .$$

This equation satisfies the initial and the single boundary condition, but it is not in a form very suitable for computation. Using, however, the trigonometric formula $2 \sin x \sin y = \cos (x - y) - \cos (x + y)$ the integral becomes

$$u = \frac{1}{\pi}\int_0^\infty f(\lambda)\, d\lambda \int_0^\infty e^{-\alpha^2 D t} [\cos \alpha (\lambda - x) - \cos \alpha (\lambda + x)]\, d\alpha .$$

Treating this exactly as before (p. 89) we obtain

$$u = \frac{1}{2\sqrt{\pi D t}} \int_0^\infty f(\lambda) \left( e^{-\frac{(\lambda - x)^2}{4Dt}} - e^{-\frac{(\lambda + x)^2}{4Dt}} \right) d\lambda . \qquad (97)$$

For the case of the muscle mentioned above where $f(x) = f(\lambda) = u_0$, after introducing the variable $\beta$ as before, and remembering the symmetry of the normal probability integral for positive and negative values of $x$, we obtain from equation (97)

$$u = \frac{2 u_0}{\sqrt{\pi}} \int_0^{\frac{x}{2\sqrt{Dt}}} e^{-\beta^2} d\beta \qquad (98)$$

which may be more simply written

$$u = u_0 \psi\left( \frac{x}{2\sqrt{Dt}} \right). \qquad (99)$$

The values of $u$ corresponding to any values of $x$, $t$ and $D$ may now readily be obtained from a table of values of the probability integral.

It frequently happens that conditions are reversed and that diffusion is into a mass of tissue, whose original initial concentration is zero, from a constant external concentration of $c$. This case may readily be dealt with, subject to the restrictions mentioned on p. 26, by noting that it is the same except for the addition of a constant term, $c$, as the imaginary case of diffusion from an internal concentration of $-c$ to an external concentration of zero. The latter case is covered by equation (98). But the constant term, $c$, may be added at will since $u = c$ is itself a solution of the diffusion equation (8). We have, therefore, as the required solution, which reduces to $u = 0$ when $t = 0$ and $u = c$ when $t = \infty$, as it should

$$u = c\left(1 - \frac{2}{\sqrt{\pi}}\int_0^{\frac{x}{2\sqrt{Dt}}} e^{-\beta^2}\, d\beta\right). \tag{100}$$

This may also be written

$$u = c\left[1 - \psi\left(\frac{x}{2\sqrt{Dt}}\right)\right]. \tag{101}$$

When the external concentration is $c$ and the initial internal concentration is not zero but some other value, $u_0$, where $c > u_0$, it is only necessary to use equation (100) for a concentration of $c - u_0$ and add $u_0$ to the result. The procedure when $c < u_0$ is equally obvious.

It will be noted that equations (100) and (101) very closely resemble (94) and (95a) except that the value of $u$ is twice as great in the former as in the latter. It is obvious therefore that the linear relation, discussed on p. 96, between the position, $x$, of some chosen concentration and the square root of the time must hold for semi-infinite as well as for infinite systems. This relation is particularly easily observed in cases of diffusion of some substance, which is either itself visible or is made visible by an indicator into a long column of agar-agar or gelatin from an external medium of constant composition (see in this connection CHABRY 1888, YEGOUNOW 1906a, b, VANZETTI 1914, AUERBACH 1924, etc.).

The case of diffusion from a layer of solution of finite thickness, originally placed at the bottom of a very tall column of solvent, has already been treated by means of an equation derived for an infinite system, though strictly speaking the system in question is semi-infinite with the boundary condition $\frac{\partial u}{\partial x} = 0$ when $x = 0$. The same problem may, if desired, be dealt with in the manner just discussed except that in order to satisfy the present boundary condition, the form of FOURIER's integral containing $\cos \alpha x$, i. e., (84b) above is used. By proceeding as before and taking advantage of the relation $2 \cos x \cos y = \cos(x + y) + \cos(x - y)$ we are led to the same equation as that originally obtained, namely, number (93).

### f) Rate of diffusion across a plane.

Among the most valuable equations relating to infinite and semi-infinite systems are those which make possible a calculation of the amount of diffusing material that crosses a given plane in any desired time. Several physiological applications of such equations will be mentioned below. The method used in obtaining them is the same as that already described for finite systems, namely, first to find the concentration gradient $\frac{\partial u}{\partial x}$ by differentiation, then to multiply this gradient by $-DAdt$, and finally to integrate the resulting expression between 0 and $t$. It will be noted that the differentiation of such equations as (94) and (100) cannot be performed in as simple a manner as is possible with a FOURIER's series. However, it is shown in the textbooks of the integral calculus that subject to certain conditions, which are here fulfilled, definite integrals of the types in question can be differentiated with respect to their variable upper limits of integration by means of the simple formula

$$\frac{d}{dz}\int_0^y f(x)\,dx = f(y)\,\frac{dy}{dz}.$$

Applying this formula to equations (94) and (100) which represent, respectively, diffusion from a semi-infinite column of solution into a semi-infinite column of water and into water which is constantly renewed at the boundary between the two media

$$\frac{\partial u}{\partial x} = -\frac{u_0}{2\sqrt{\pi D t}}\,e^{-\frac{x^2}{4Dt}} \qquad (102\,\mathrm{a})$$

and

$$\frac{\partial u}{\partial x} = -\frac{u_0}{\sqrt{\pi D t}}\,e^{-\frac{x^2}{4Dt}}. \qquad (102\,\mathrm{b})$$

The plane of most interest is that represented by $x = 0$. Substituting this value of $x$, multiplying by $-DAdt$ and integrating between $t = 0$ and $t = t$, we obtain for the two cases, in the order mentioned above, the very important equations

$$Q_{0,t} = u_0 A\sqrt{\frac{Dt}{\pi}} \quad \text{and} \qquad (103)$$

$$Q_{0,t} = 2\,u_0 A\sqrt{\frac{Dt}{\pi}}. \qquad (104)$$

It is interesting to note that the amount of substance that passes across the plane $x = 0$ is exactly twice as great when the concentration at this level is kept at zero as when it is allowed to change in accordance with the ordinary laws of diffusion. This relation, however, is what would be expected from a consideration of the theory of random molecular movements. When in diffusion a molecule reaches the plane, $x = 0$, the chances that it will go forward or backward are each 1/2. But, if by stirring or some other method of removal, all the molecules that reach the plane are carried forward, then the number actually crossing this level must be exactly twice as great as before.

Equation (103) can be applied to infinite systems in which stirring or mixing is absent, and in which $D$ has the same, or nearly the same, value in the two halves of the system. This would be the case in diffusion from one aqueous medium to another, and even approximately so in diffusion from an agar-agar or gelatin gel of not too great concentration to water. The necessary conditions would, however, no longer hold in diffusion from muscle to an unstirred aqueous medium since the effective diffusion coefficients in the two media are too different. Such cases as this could perhaps be dealt with by the more complicated method of treatment of systems composed of two dissimilar regions given by FÜRTH (1927c, p. 192). It is far simpler, however, to use a stirred external medium and to apply equation (104).

Equation (104) has been used for physiological purposes by a number of workers (STELLA 1928, EGGLETON and HILL 1928, HILL 1928, ADOLPH 1931). The principle involved is that if diffusion occurs for a relatively short time between a large mass of tissue such as an entire muscle and an external stirred medium of practically constant composition, the innermost parts of the tissue are not appreciably affected by the diffusion process, and the case may with very little error be treated as one involving a semi-infinite solid with a plane surface equal in extent to the entire surface of the tissue. As the time becomes longer, the assumptions that the body is semi-infinite and that diffusion in it occurs in only one dimension at right angles to a plane surface both become less and less tenable. Experience shows, however, that the times within which equation (104) is approximately true are entirely suitable for experimental work.

As an example of the application of the method, the following results reported by EGGLETON, EGGLETON and HILL (1928) may be cited as typical. A pair of frog's legs was skinned and stimulated in nitrogen. It was then plunged rapidly into 20 c.c. of oxygen-free RINGER's solution. After a time, $t_1$, it was placed in a second fresh 20 c.c. sample, and after a further time, $t_2$, into a third similar sample. Analyses were made of the solutions to determine the amount ($Q$) of lactate lost by the muscles during each interval. As long as equation (104) is applicable to the process, there should be a linear relation between $Q$ and the square root of the time. In the present case the relations observed were

$$t_1 = 12.5 \text{ minutes} \qquad t_2 = 27.5 \qquad t_3 = 49$$
$$Q_1 = 0.59 \text{ mg.} \qquad Q_2 = 0.86 \qquad Q_3 = 1.18$$
$$Q_1 \sqrt{t_1} = 0.167 \qquad Q_2 \sqrt{t_2} = 0.164 \qquad Q_3 \sqrt{t_3} = 0.169.$$

The agreement is excellent, indicating that it is permissible to apply the equation in question to the calculation of diffusion coefficients. For this purpose the other necessary data are that the amount of lactate found in the muscles at the end of the experiment plus that which had escaped was 4.9 mg., distributed in 4.7 c.c. of muscles, indicating therefore a value of $u_0$ of 1.05; and that the area of the muscles was 17 cm.

Introducing these figures into equation (104) a value of $D$ of $6.9 \times 10^{-5}$ cm.²/minute is obtained. This value is much lower than that found by the same investigators for the same substance by essentially the same method for agar jelly ($6.6 \times 10^{-4}$) or than those already known for related substances in water. Furthermore, the value of the coefficient changes greatly with the condition of the muscle, decreasing during extreme fatigue to $5 \times 10^{-4}$, but increasing in heat rigor to a value of approximately twice that associated with normal muscle. These results and the somewhat similar one reported by STELLA (1928) can best be accounted for by assuming that the diffusion in muscle of substances such as lactates, phosphates, etc. takes place chiefly in the lymph spaces rather than across the cell membranes. The volume of these spaces is evidently subject to considerable decrease in fatigue, while in heat rigor, on the other hand, injury of the individual muscle cells probably increases their permeability and in this way increases the value of the diffusion coefficient.

It is of interest in this connection to compare the diffusion coefficients given by KROGH (1919a) for oxygen in muscle and in water, respectively (p. 67). It will be noted that the former is not greatly less than the latter; this fact would seem to indicate a much greater permeability of muscle cells to oxygen than to lactates, phosphates, etc. — which is what would be expected on other grounds. It will be noted that diffusion coefficients for muscle, because of the complexity of the structure of this tissue, are not in all respects comparable to those for water and for other homogeneous media, since they are based merely upon an over-all effect for the tissue as a whole. But since they describe accurately and quantitatively the course of diffusion in such a tissue they can be used with considerable confidence for making the calculations which from time to time become necessary in physiological work.

Equation (103), though not so useful for physiological purposes as (104), is nevertheless frequently of much value in dealing in an approximate manner with diffusion processes in finite homogeneous systems. For example, under those conditions where the infinite series in such an equation as (32) converges too slowly for convenience, (103) may with little error be substituted, and the necessary calculations may then be made with ease. Thus, suppose that diffusion between the two halves of the system represented in equation (32) has taken place for a relatively short time. Under these conditions the regions farthest from the original boundary of the solution and the solvent are so little affected that the system, for practical purposes, may be considered to be infinite. For such a case the amount of substance, $Q$, that has passed from one half of the vessel to the other is given by equation (103) and we have

$$\frac{Q_1 - Q_2}{Q_1 + Q_2} = \left(\frac{u_0 H A}{2} - 2 u_0 A \sqrt{\frac{D t}{\pi}}\right) \Big/ \frac{u_0 H A}{2} = 1 - \frac{4}{\sqrt{\pi}} \sqrt{\frac{D t}{H^2}}. \quad (105)$$

It was, indeed, by means of this very simple relation that the values in the second column of the table on p. 45 for small values of $Dt/H^2$ were calculated.

The useful approximation used by INGHAM and by ROUGHTON (1932) (see p. 52) may be derived in exactly the same way by means of equation (104). Instead of representing the degree of saturation of the finite system with one open boundary in which they were interested by means of equation (43) when $\frac{Dt}{4H^2}$ is small, the degree of saturation may be obtained simply by dividing $Q$, as given by equation (104), by the volume of the system. This leads at once to the desired result, namely

$$\frac{\bar{u}}{c} = \frac{2}{H}\sqrt{\frac{Dt}{\pi}}. \qquad (106)$$

## 11. Brownian movement and diffusion.

The discussion of diffusion processes in the preceding sections has been based upon FICK's law. For most purposes, this method of treatment is the best one available, but it should not be forgotten that there is another entirely different method of approach, which historically has played an important part in the development of a number of modern concepts in physics and physical chemistry, and which has certain advantages of its own as a means of dealing with problems of diffusion. The alternative method is to consider statistically the behavior of the individual diffusing molecules, using as a visible model of their behavior the Brownian movement of small particles suspended in a liquid medium. This method is evidently very different from that based on FICK's law, which makes no assumptions as to the exact molecular constitution of solutions, and which is therefore equally applicable to the flow of heat in solid bodies, in which movements of translation of discrete particles is not involved.

The subject of Brownian movement is very adequately dealt with in a number of special monographs (PERRIN 1909, 1914, SVEDBERG 1912, DE HAAS-LORENZ 1913, FÜRTH 1931 b, etc.) to which the reader is referred for historical and other details which must be omitted here for lack of space. It was not until three quarters of a century after its description and somewhat crude experimental treatment by BROWN (1828) — it had been noticed and commented upon by other microscopists even earlier — that Brownian movement became more than an object of scientific curiosity, and received the serious attention from physicists that it deserved. The first person to deal with the subject theoretically in an entirely satisfactory manner and to give it a definite quantitative relation to other physical phenomena was EINSTEIN (1905 and later), whose more important papers in the field have been collected in a single small volume (EINSTEIN 1922). About the same time, and entirely independently, v. SMOLUCHOWSKI (1906 and later) obtained very similar results, which have likewise been collected in a convenient volume (v. SMOLUCHOWSKI 1923). The theoretical treatment of the subject by these two investigators and its experimental study, particularly by PERRIN (1909), SVEDBERG (1912) and their respective collaborators, forms one of the most important chapters in the history of modern physics.

### a) The mean squared displacement.

While the study of Brownian movement may be said to have definitely established the truth of the molecular and the kinetic theories of the constitution of matter, and while it has received practical applications in as diverse fields as that of the measurement of the charge carried by an electron and of the nature of the factors that determine the limits of sensitivity of a galvanometer, it will be considered here only in its relation to diffusion processes. This consideration may appropriately begin with a simple statistical treatment of the behavior of a large number of solute molecules or of visible particles showing Brownian movement.

In the first place, the hypothesis may be made — and its truth is confirmed by all the available experimental evidence — that a freely suspended particle or solute molecule, surrounded on all sides by solvent molecules and other particles or molecules similar to itself, and in the state of intense agitation demanded by the kinetic theory, will be equally likely to move in any direction. Since a study of the movement of such a particle in three dimensions presents both experimental and theoretical difficulties, the problem may be simplified by restricting it merely to the components of the actual movements which lie in a single dimension, in which the distance of the particle from some chosen plane may be designated by $x$. In general, every one of the continual and countless collisions which the particle suffers with the surrounding molecules will result in some change in $x$, and these changes are equally likely to be in the positive or the negative direction. Furthermore, because of the entire symmetry of the system, whatever the total effect of some given large number of collisions might be, such an effect, if repeatedly observed, would be found to occur just as often in the one direction as in the other.

When the behavior of a single visible particle showing Brownian movement is studied under the microscope, it is obviously impossible to detect the effects of single collisions; all that can be done is to observe the total effect in a time measured in, say, minutes, of an inconceivably great number of collisions. Experience shows that if a single particle be observed at two times separated by a convenient interval, it undergoes during that interval some measurable displacement in the $x$-direction.

This displacement may be called $\Delta_x$. How is $\Delta_x$ related to time? This question is very clearly discussed by DE HAAS-LORENZ (1913) in a manner which may be roughly reproduced as follows. Let the total number of collisions that have resulted in a given displacement be represented by $n$, and the separate displacements following each collision by $s_1$, $s_2$, $s_3$, etc. Then,

$$\Delta_x = s_1 + s_2 + s_3 + \ldots \tag{107}$$

where the different values of $s$, in general, differ in magnitude and may be either positive or negative in sign. Since a single observation on one particle gives no possible means of predicting the behavior of other particles, or of the same particle at other times, a statistical study of many observed displacements is necessary; and it will be of value in proportion to the number of separate observations on which it is based.

It is obviously useless to attempt by repeated observations to find an average value of $\Delta_x$ — generally represented by $\overline{\Delta_x}$ — since both theory and experience tell us that in a completely symmetrical system this value must approach zero as the number of observations is indefinitely increased. If, however, the effect of the magnitude of each observed value of $\Delta_x$ be retained but that of its sign be eliminated by squaring the values in question, an average value of the squared displacements, i. e., $\overline{\Delta_x^2}$, can then be obtained which has great statistical significance, being in fact, identical with the square of the measure of variability, $\sigma$, called the standard deviation by statisticians. But this value, by equation (107), must be

$$\overline{\Delta_x^2} = \overline{s_1^2} + \overline{s_2^2} + \overline{s_3^2} + \ldots + 2\,\overline{s_1 s_2} + 2\,\overline{s_1 s_3} + \ldots + 2\,\overline{s_2 s_3} + \ldots$$

Of the separate terms used to obtain the averages on the right-hand side of the last equation, those having the form of squares are all positive; the remainder are equally likely to be positive and negative, and their averages must therefore approach zero as the number of observations is indefinitely increased. Furthermore, for a very large number of observations, $\overline{s_1^2} = \overline{s_2^2} = \overline{s_3^2} = \overline{s^2}$, etc., and we therefore have

$$\overline{\Delta_x^2} = n\,\overline{s^2}.$$

But $n$, the total number of collisions, must be equal to the total time, $t$, divided by the average time between collisions $\tau$, and the important relation is obtained

$$\overline{\Delta_x^2} = \frac{\overline{s^2}}{\tau}\,t = k\,t. \tag{108}$$

In other words, on the average, there must be a proportionality, not beween time and displacement in a given direction, but between time and the square of the displacement. But this relation has already been encountered in a very different form in every equation in the preceding sections having to do with finite systems in which the exponent $-\frac{n^2 \pi^2 D t}{H^2}$ is involved; obviously in all such equations constancy of $n$ and $Q$ demands a constancy of the ratio $t/H^2$. Similarly, in all the equations pertaining to infinite systems in which the relation $\frac{x}{2\sqrt{Dt}}$ in any way appears, the same principle is involved. There is, in fact, no more characteristic property of diffusion processes than this relation of distance to the square root of the time. (See also in this connection p. 96.)

The same treatment as that just applied to one dimension may readily be extended to two and three dimensions. Thus, let the displacement in the $xy$ plane be designated by $\Delta_v$. Then $\Delta_v^2 = \Delta_x^2 + \Delta_y^2$. But for $\Delta_y^2$ there can be obtained exactly the same sort of expression as for $\Delta_x^2$. Therefore

$$\overline{\Delta_v^2} = 2\,n\,\overline{s^2}$$

and by the same reasoning for three dimensions

$$\overline{\Delta_r^2} = 3\,n\,\overline{s^2}.$$

It follows that whatever the number of dimensions, the mean squared displacement in that number of dimensions is proportional to the time. In other words, on the average, the square of the total distance in a straight line that a particle travels from its starting point is also proportional, though the factor of proportionality different for the cases of one and of three dimensions, to the time.

### b) Brownian movement and probability.

It is possible by an extension of the simple type of reasoning so far employed to draw more detailed deductions about the statistical behavior of particles in Brownian movement. Suppose that within a time, $t$, long enough to permit the making of ordinary observations, a particle be subjected to $n$ collisions. Let a distance $s$ be defined by dividing the total distance travelled by the particle both forward and backward in the $x$-direction by $n$. As determined in this way, for times measured in hours, minutes, or even seconds, $s$ will have a highly reproducible value and we may adopt the convenient fiction of supposing that the progress of the molecule has taken place in $n$ equal steps each of length $s$, some forward and others backward. Now although the *total* distance travelled by a particle in one dimension in a reasonably long time will differ only negligibly from experiment to experiment, the distance of its observed position from the starting point will be highly variable, since forward and backward steps of length $s$ can occur in all possible proportions. From the conditions of symmetry in the system we know that the chances of forward and backward steps are each equal to 1/2. But this is exactly the situation from which numerous derivations of the so-called normal probability curve take their origin. Omitting this derivation and merely making use of the well-known end result, we have for a distribution in which the standard deviation is represented by $\sigma$, a probability that a given deviation shall fall between the limits $\pm x$

$$P_{\pm x} = \frac{2}{\sigma\sqrt{2\pi}} \int_0^x e^{-\frac{x^2}{2\sigma^2}} dx.$$

By a suitable transformation this becomes

$$P_{\pm x} = \frac{2}{\sqrt{\pi}} \int_0^{\frac{x}{\sigma\sqrt{2}}} e^{-x^2} dx \qquad (109)$$

which is the form found in most tables of the normal probability integral. In the present case $\sigma$ is evidently equal to $\sqrt{\Delta_x^2}$, which can readily be determined from a sufficient number of observations.

An experimental test of the relation of Brownian movement to the probability integral was made by PERRIN (1909) from data obtained by CHAUDESAIGUES. In the experiments in question, particles of gamboge, of as nearly as possible of the same radius ($0.212\,\mu$) were observed under the microscope, and the components of their displacements in a single dimension in a given time were measured. In one series, the following data were obtained from observations on 205 particles.

| Displacement in micra | Number Observed | Number Calculated |
|---|---|---|
| 0 — 1.7 | 48 | 44 |
| 1.7— 3.4 | 38 | 40 |
| 3.4— 5.1 | 36 | 35 |
| 5.1— 6.8 | 29 | 28 |
| 6.8— 8.5 | 16 | 21 |
| 8.5—10.2 | 15 | 15 |
| 10.2—11.9 | 8 | 10 |
| 11.9—13.6 | 7 | 6 (5) |
| 13.6—15.3 | 4 | 4 |
| 15.3—17.0 | 4 | 2 |
| Total | 205 | 205 |

To calculate the theoretical number of displacements of any given magnitude, $\sigma$ (i. e., $\sqrt{\Delta_x^2}$) is first obtained from the data in the first and second columns. Since the individual displacements are not given exactly but in groups falling between two limits, the arithmetical mean of these limits may be used for the calculation. The work is simplified by measuring $\Delta_x$ not in micra but in units of $1.7\,\mu$. Expressed in this unit rather than in micra, $\sigma^2 = 13.84$. Introducing this value into equation (109) to obtain the probabilities of displacements not greater than 1, 2, 3, ... respectively, multiplying these probabilities by the total number 205, and taking the differences between the values successively calculated in this way, the figures shown in the last column are obtained. The figure in parenthesis is that given by PERRIN, but the one substituted for it seems to be preferable. The agreement between theory and observation is extremely good.

**c) The relation between displacement and the diffusion coefficient.**

It will be noted that by means of equation (109) the manner of distribution in one dimension of a number of particles originally arranged in a very thin layer may be calculated for any desired value of $t$ if $\sigma$ be known. But another equation [(90), p. 90] has already been derived in a very different manner for exactly the same case, and differs from the present one only in containing as the upper limit of integration $2\sqrt{Dt}$ in place of $\sigma\sqrt{2}$. It follows, therefore, if both equations be true, that since $\sigma^2 = \overline{\Delta_x^2}$ we may write the equation

$$\overline{\Delta_x^2} = 2\,D\,t. \tag{110}$$

This same relation may, indeed, be obtained from equation (87) in a different and a somewhat preferable way. By means of the latter equation, starting with a number of particles, $N$, arranged in a plane, which is taken as the origin for the measurement of $x$, the number of particles that at any time $t$ have travelled a distance $x$ measured perpendicularly to this plane (i. e., the number that at the time in question lie between the levels $x$ and $x + dx$) is

$$n_x = \frac{N}{2\sqrt{\pi D t}}\, e^{-\frac{x^2}{4Dt}}\, d\,x\,.$$

If now this number for each distance be multiplied by the square of that distance, if the sum of all possible terms of this sort be obtained by integration, and if then, finally, this sum be divided by the total number $N$, the result will be the mean of the squared deviations, i. e., $\overline{\Delta_x^2}$. We have therefore

$$\overline{\Delta_x^2} = \frac{1}{2\sqrt{\pi D t}} \int_{-\infty}^{\infty} x^2\, e^{-\frac{x^2}{4Dt}}\, d\,x\,.$$

This integral may be simplified by introducing a new variable $z = \frac{x^2}{4Dt}$, giving

$$\overline{\Delta_x^2} = \frac{4Dt}{\sqrt{\pi}} \int_0^{\infty} \sqrt{z}\, e^{-z}\, d\,z\,.$$

Integrating by parts

$$\overline{\Delta_x^2} = \frac{4Dt}{\sqrt{\pi}} \left[-\sqrt{z}\, e^{-z}\right]_0^{\infty} + \frac{4Dt}{\sqrt{\pi}} \int_0^{\infty} e^{-z} \frac{d\,z}{2\sqrt{z}}\,.$$

The expression enclosed in square brackets vanishes and the integral which remains, on again substituting a new variable $u = \sqrt{z}$, becomes

$$\overline{\Delta_x^2} = \frac{4Dt}{\sqrt{\pi}} \int_0^{\infty} e^{-u^2}\, d\,u = 2\,D\,t\,.$$

We are therefore led in this way to exactly the same equation as that already given above (110) which was in fact originally derived from first principles by EINSTEIN (1905, 1908).

Another example of the same important relation between the diffusion coefficient and the mean squared displacement of individual molecules is obtained by comparing the SUTHERLAND-EINSTEIN equation [(3), p. 12] with a similar one due to LANGEVIN (1908), the derivation of which presents several points of interest. Consider the displacement parallel to the $x$-axis of a particle in Brownian movement in consequence of the impacts of the surrounding molecules. Let $X$ be the component of the forces due to these impacts which acts in the $x$-direction. The effective force acting on the particle and bringing about its displacement is equal to its mass, $m$, multiplied by its acceleration, $\frac{d^2 x}{d t^2}$. This effective force is evidently also equal to $X$ minus the frictional resistance at unit velocity, $f$, encountered by the particle multiplied by its velocity, $\frac{d x}{d t}$. We therefore have

$$m \frac{d^2 x}{d t^2} = -f \frac{d x}{d t} + X.$$

Multiplying both sides of the equation by $x$ it becomes

$$m x \frac{d^2 x}{d t^2} = -f x \frac{d x}{d t} + X x.$$

A new variable, $x^2$, may now be introduced by the routine methods of the differential calculus, giving for any single particle

$$\frac{m}{2} \frac{d^2 (x^2)}{d t^2} - m \left(\frac{d x}{d t}\right)^2 = -\frac{f}{2} \frac{d (x^2)}{d t} + X x.$$

If analogous equations be employed for a large number of particles and all be added and averaged, an expression involving $\overline{x^2}$ is obtained. At the same time the average of the $X x$ terms disappears, since positive and negative values of $x$ tend to balance each other. In place of $m\left(\frac{d x}{d t}\right)^2$ may be written $mv^2$, which, it will be noted, is twice the value of the kinetic energy of the particle parallel to the $x$-axis. Now from the kinetic theory of gases, the total kinetic energy of a particle taking part in the random molecular movement in a system of this sort is $\frac{3}{2} \frac{R T}{N}$. This energy is, on the average, equally distributed in three directions at right angles to one another, and its magnitude, therefore, in the one dimension under consideration must be equal to one-third of the total value. The equation may therefore be written

$$\frac{m}{2} \frac{d^2 (\overline{x^2})}{d t^2} - \frac{R T}{N} = -\frac{f}{2} \frac{d (\overline{x^2})}{d t}.$$

Introducing a new variable, $p = \frac{d (\overline{x^2})}{d t}$, the equation becomes a simple linear differential equation of the first order, namely,

$$\frac{d p}{d t} + \frac{f}{m} p = \frac{2 R T}{N m}$$

whose solution, obtained by the use of the integrating factor $e^{\frac{f}{m}t}$, followed by the restoration of the original variable is

$$\frac{d\,(\overline{x^2})}{d\,t} = \frac{2\,R\,T}{N f} + C_1 e^{-\frac{f}{m}t}$$

When $t$, according to LANGEVIN, is of an order of magnitude that can be measured, the last term may be neglected. Dropping it, and again integrating, the result is

$$\overline{x_2} = \frac{2\,R\,T}{N f}\,t + C_2\,.$$

But when $t = 0$, $x = x_0$ and the equation becomes

$$\overline{x^2} - \overline{x_0^2} = \frac{2\,R\,T}{N f}\,t\,.$$

Now since $x = x_0 + \Delta_x$, $x^2 = x_0^2 + 2\,x_0\,\Delta_x + \Delta_x^2$; and on taking average values, the term containing $x$, which is equally likely to be positive and negative, tends to vanish. The equation therefore becomes

$$\overline{\Delta_x^2} = \frac{2\,R\,T}{N f}\,t = \frac{R\,T}{N} \cdot \frac{1}{3\,\pi\,\eta\,r}\,t\,. \tag{111}$$

On comparing this equation (into which the value of $f$ obtained by means of STOKES' law either may or may not be introduced) with the SUTHERLAND-EINSTEIN equation, the relation shown in equation (110) again appears.

Equation (111) has been tested in different ways. The direct relation between $\overline{\Delta_x^2}$ and $t$ which it demands was shown in fact to exist by CHAUDESAIGUES (1908), and more fully by CHAUDESAIGUES and PERRIN in the experiment cited above (p. 106). CHAUDESAIGUES also demonstrated an approximately inverse proportionality between $\overline{\Delta_x^2}$ and the radius of the particles, and between $\overline{\Delta_x^2}$ and the viscosity of the medium, other factors in each case being equal. SEDDIG (1908) had also found that a change in temperature, which affects both $T$ and $\eta$ in equation (111), has approximately the predicted effect, though there was in his results a disagreement of about six per cent between theory and observation which he attributed to an experimental error. The most conclusive test of the equation, however, was furnished by PERRIN who calculated $N$ for two different types of suspensions (gamboge and mastic) for several different temperatures, for viscosities, ranging from that of water to one increased 125 times by the addition of glycerol, and for radii whose extreme values were 0.212 $\mu$ and 5.5 $\mu$ respectively. In all cases, not only did the calculated values of $N$ show a good agreement among themselves, but these values were fairly close to those obtained by entirely independent methods.

PERRIN (1909) has summarized some of his chief results and SVEDBERG (1912, p. 116) has made a more complete tabulation of them. The value of $N$ indicated by all the experiments of this type is, according to PERRIN, $6.85 \times 10^{23}$, which may be compared with the value of $6.88 \times 10^{23}$ obtained by the same investigator by the sedimentation equilibrium method (p. 128). Though this value is of the right order of magnitude, it is slightly higher than the accepted value of $6.06 \times 10^{23}$. More recently SHAXBY (1923, 1925), thinking that better results would be obtained by the use of particles more perfectly wetted by water than those employed by PERRIN, made similar studies with spherical bacterial cells (cocci) and obtained the very satisfactory values of $N$ with Staphylococcus albus of $6.08 \times 10^{23}$ and with S. aureus of $6.22 \times 10^{23}$.

### d) Behavior of a finite number of particles.

Though, in general, the normal probability distribution, on which much of the foregoing discussion is based, provides a highly satisfactory method of dealing with the behavior of individual molecules and particles in Brownian movement, it has one limitation, namely, that it describes the theoretical behavior of an infinite number of units, while the number involved in any actual case, though enormously large, is nevertheless finite. The equation of the normal probability curve demands that in any time, however small, *some* diffusing material shall find its way to any distance, however great. This is obviously impossible for a finite number of molecules or particles, and it is not permissible, therefore, to employ this distribution in its usual form to calculate the extreme displacements to be expected in any given system. FRANK (1918), however, has shown how this difficulty may be dealt with theoretically, and SITTE (1931) has made actual observations which are in excellent agreement with the theory, of which the following is a very brief summary.

By equation (90) or by equation (109) after the substitution of $\sqrt{2Dt}$ for $\sigma$, the number of molecules, $n$, that at the time $t$ will on the average lie between the limits $\pm x$ will be

$$n = N\psi\left(\frac{x}{2\sqrt{Dt}}\right)$$

where $\psi(z)$ represents the probability integral for the argument $z$. The number that will exceed this distance is obtained by subtracting the right-hand side of this equation from $N$. When only a single molecule exceeds the distance $x$, we may take $x$ as the boundary of the diffusing swarm, and the distance in question is found from the relation

$$\frac{1}{N} = 1 - \psi\left(\frac{x}{2\sqrt{Dt}}\right) \quad \text{or} \quad x = 2\sqrt{Dt}\,\psi^{-1}\left(1 - \frac{1}{N}\right) \qquad (112)$$

where $\psi^{-1}(x)$ bears the same relation to $\psi(x)$ that $\sin^{-1}x$ does to $\sin x$. For small values of $N$, $x$ may readily be calculated from ordinary tables of the probability integral. When $N$ is very large we may use the series

$$1 - \psi(z) = \frac{e^{-z^2}}{z\sqrt{\pi}}\left(1 - \frac{1}{2z^2} + \frac{1.3}{(2z^2)^2} - \ldots\right)$$

where $z = \frac{x}{2\sqrt{Dt}}$. If $N$ be extremely large and $x$ therefore sufficiently great, so that approximately

$$\frac{1}{N} = \frac{e^{-\frac{x^2}{4Dt}}}{\frac{x\sqrt{\pi}}{2\sqrt{Dt}}}$$

we may, after taking logarithms of both sides, and remembering that $\ln\left(\frac{x\sqrt{\pi}}{2\sqrt{Dt}}\right)$ is negligible in comparison with $\ln N$, finally obtain the relation

$$x = 2\sqrt{Dt}\sqrt{\ln N}. \tag{113}$$

That is to say, the distance, $x$, travelled by the „head“ of the swarm of molecules in a given time, say 1 second, increases without limit with increasing values of $N$; but the rate of increase is surprisingly slow. Thus, to double the value of $x$, in the region to which equation (113) applies, it would be necessary to increase $N$ in the proportion given by $\frac{\ln N_2}{\ln N_1} = 4$ i. e., $N_2 = N_1^4$; to treble the value of $x$, it would be necessary to take $N_2 = N_1^9$, and so on.

### e) The mean time of two-sided first passage.

It was mentioned in the introduction that Brownian movement has recently been applied to a study of some of the physical characteristics of protoplasm. One such study was made by Becking, Bakhuyzen and Hotelling (1928) who measured the displacements at fixed intervals of particles 0.4 $\mu$ in diameter in the parietal protoplasm of *Spirogyra*, and applied to these displacements equation (111). It was found that a value of $\eta$ could be determined in this way which varied in a characteristic manner with temperature and could be used to throw light on the physical properties of the protoplasm; but it was emphasized that this value could not be treated as the true viscosity of a liquid, since protoplasm is not a simple liquid. Indeed, from various discrepancies between the theoretical and observed behavior of the particles, the conclusion could be drawn that Brownian movement in the protoplasm of *Spirogyra* is subject to certain hindrances which are probably of a structural nature.

The method used by Becking et al., which consists in measuring the horizontal displacements of a sufficient number of particles in a given time, while theoretically sound, is somewhat more troublesome to carry out than another method due to FÜRTH (1917) and called by him the method of the mean two-sided time of first passage. The principle of the method, which promises to become of great usefulness in biological investigations (see PEKAREK 1930 and following papers), is to introduce into the field of a microscope a number of parallel and equally spaced lines at a distance $b$ from one another. The time when a particle crosses any one of these lines is noted, and also that when it crosses the next line on either side. The observation is then repeated many times, either with the same or with a different particle, and the average time of passage of the next line in either direction is obtained. It will be shown below that this average time bears a very simple relation to the diffusion coefficient of the particle in question. The reason for not selecting for observation a single one of the two possible lines that might be crossed is that since, theoretically, half of all the particles must move in one direction and half in another, the average time for crossing the next line in a previously specified direction would be infinite. It is shown by FÜRTH that even with a single line, an equation can be derived which relates the diffusion coefficient to the average of all times of passage not greater than some selected value. This method, however, is inferior to the two-line method, which because of its probable practical importance in biological research may now be discussed in greater detail.

Imagine a particle at the moment it leaves in either direction a given line represented by $x = b$. The two adjacent lines are taken as $x = 0$ and $x = 2b$, respectively, $b$ being the distance between any two lines. What is the probability that in the interval between $t$ and $t + dt$ the particle will cross *either* of the other two lines? The problem may be considered analogous to that of diffusion in a slab of thickness $2b$ where, since a particle disappears from consideration as soon as it crosses one of the boundaries, the boundary conditions may be taken as $u = 0$ when $x = 0$ and $u = 0$ when $x = 2b$. In order to make the quantities crossing the two boundaries assume the form of probabilities, the initial quantity of diffusing substance is taken as unity but the unit employed may if desired be such as to make it represent an inconceivably great number of molecules. These molecules at the beginning of the time under consideration are contained in a layer of infinitesimal thickness, so the initial condition is: when $t = 0$, $u = 1$ for $x = b$ and $u = 0$ for all other values of $x$.

An equation [number (23)] has already been derived which by merely replacing $H$ by $2b$ fits this particular case. After introduction of the appropriate initial condition, this equation becomes

$$u = \frac{1}{b} \sum_{n=1}^{n=\infty} \left( e^{-\frac{n^2 \pi^2 D t}{4 b^2}} \sin \frac{n \pi x}{2b} \sin \frac{n \pi}{2} \right). \qquad (114)$$

The amount of material, $dQ$, that will cross unit area of the two boundaries by FICK's law is

$$D\left[\left(\frac{\partial u}{\partial x}\right)_{x=0} - \left(\frac{\partial u}{\partial x}\right)_{x=2b}\right] dt \tag{115}$$

where the significance of the different algebraic signs of the two gradients is that the movement across the two boundaries is in opposite directions. Performing on equation (114) the operations indicated in (115), we have

$$\frac{dQ}{dt} = \frac{D\pi}{2b^2}\sum_{n=1}^{n=\infty}\left[e^{-\frac{n^2\pi^2 Dt}{4b^2}}\, n \sin\frac{n\pi}{2}\,(1-\cos n\pi)\right].$$

Remembering that $1-\cos x = 2\sin^2\frac{x}{2}$ and that $\sin^3\frac{n\pi}{2} = \sin\frac{n\pi}{2}$ this reduces to

$$dQ = \frac{D\pi}{b^2}\sum_{n=1}^{n=\infty}\left(n\sin\frac{n\pi}{2}\, e^{-\frac{n^2\pi^2 Dt}{4b^2}}\right) dt. \tag{116}$$

[On integration of (116) between $t=0$ and $t=\infty$, $Q=1$ as it should.] Since a unit amount of the diffusing substance was used in deriving equation (114), the right-hand side of the last equation gives the probability that a single molecule will cross one of the two boundaries between the times $t$ and $t+dt$. If now each time be multiplied by its respective probability, if the sum of all the products be obtained by integration for the entire period from $t=0$ to $t=\infty$ and if this sum then be divided by the total of all the probabilities, i.e., by unity, the value obtained is that of the time which on the average in a large number of observations would be required for a molecule to cross a line at a distance $b$ in either direction from its starting point. Calling this time $\vartheta$, we have

$$\vartheta = \frac{\pi D}{b^2}\sum_{n=1}^{n=\infty}\left(n\sin\frac{n\pi}{2}\int_0^\infty t\, e^{-\frac{n^2\pi^2 Dt}{4b^2}}\, dt\right)$$

$$= \frac{\pi D}{b^2}\cdot\frac{16\, b^4}{n^4\pi^4 D^2}\sum_{n=1}^{n=\infty} n\sin\frac{n\pi}{2}.$$

Remembering that $\frac{\pi^3}{32} = 1 - \frac{1}{3^3} + \frac{1}{5^3} - \ldots$, the extremely simple and useful relation is finally obtained

$$\vartheta = \frac{b^2}{2D}. \tag{117}$$

This equation immediately suggests (110) on p. 107 which has a very similar form. Both show a simple proportionality, into which the diffusion coefficient enters in the same way, between time and a distance squared. But whereas equation (110) gives the average squared displacement in a fixed time, the present one gives the average time required for a fixed displacement without regard to sign. As an example of the reproducibility and reliability of the method the following observations by FÜRTH (1917) may be cited. A mastic suspension was used and the average time was determined which a particle, starting from a given line, required to cross the $b$ th line *either* to the right or the left. The value of $\frac{b^2}{\vartheta}$ shows a very satisfactory constancy and that of $D$, determined by means of equation (117), is also in good agreement with that obtained by another independent method.

| $b$ | $\vartheta$ | $\frac{b^2}{\vartheta}$ |
|---|---|---|
| 1 | 3.73 | 0.267 |
| 2 | 15.09 | 0.265 |
| 3 | 33.39 | 0.269 $D = 0.133$ |
| 4 | 59.17 | 0.269 |
| 5 | 94.09 | 0.265 |
| 6 | 138.57 | 0.259 |

## 12. Diffusion in cylinders.

The cases of diffusion so far considered have all been uni-dimensional. There are, however, certain others of physiological importance which do not fall under any of the headings so far discussed, and which require other methods for their mathematical treatment. One of the most frequently encountered is that involving symmetrical diffusion into or out of a long cylinder such as a nerve, an individual muscle fiber or a cylindrical muscle. Where the relative depth affected by diffusion is not too great, the system may frequently be dealt with by the approximate method already described in section 10 f, but with cylinders of small diameter, or with long times, or both, this simplification is not permissible, and equations for diffusion in one dimension no longer suffice. The most general case of diffusion into a cylinder is a problem of diffusion in three dimensions; this case has been treated by WILLIAMSON and ADAMS (1919). It usually happens, however, in structures of physiological importance such as nerves, individual muscle cells, etc., that the cylinder in question is so long in comparison to its diameter that diffusion into its ends is insignificant and may be neglected. The problem then becomes one of diffusion in two dimensions instead of three, with a consequent simplification of its mathematical treatment.

The general diffusion equation expressed in cylindrical coordinates has already been given above [(9), p. 23]. When the distribution of the material is such that $u$ is a function of $r$ alone, i. e., when there is no diffusion in the direction measured by $z$ and when the distribution of material about the $z$-axis is at all times symmetrical, then equation (9) assumes the much simpler form

$$\frac{\partial u}{\partial t} = D\left(\frac{\partial^2 u}{\partial r^2} + \frac{1}{r}\frac{\partial u}{\partial r}\right). \tag{118}$$

The solution of this differential equation requires the use of BESSEL's functions, which are less familiar to most biologists than the trigonometric and exponential functions so far employed, and a very brief description of some of their elementary properties may, therefore, be desirable before proceeding with the solution of the equation. By definition, a BESSEL's function of order $n$, when $n$ is an integer—this being the only case that concerns us here—may be written

$$J_n(x) = \frac{x^n}{2^n n!}\left[1 - \frac{x^2}{2(2n+2)} + \frac{x^4}{2\cdot 4(2n+2)(2n+4)} - \ldots\right].$$

For the only two orders which will be required in the present discussion, namely, zero and one, we have, remembering that both 1! and 0! are equal to unity

$$J_0(x) = 1 - \frac{x^2}{2^2} + \frac{x^4}{2^2\cdot 4^2} - \ldots \quad \text{and} \quad J_1(x) = \frac{x}{2} - \frac{x^3}{2^2\cdot 4} + \frac{x^5}{2^2\cdot 4^2\cdot 6} - \ldots$$

By differentiating or integrating, term by term, similar series in which $x$ has been replaced by $kx$, where $k$ is a constant, we obtain formulae which will be needed later, namely

$$\frac{d(J_0(kx))}{dx} = -kJ_1(kx) \quad \text{and} \quad \int_0^a xJ_0(kx)\,dx = \frac{a}{k}J_1(ka). \qquad (119)$$

It is characteristic of BESSEL's functions that when plotted graphically, taking $x$ as the independent variable, the graph crosses the $x$-axis an infinite number of times. At each point of crossing we have a root of the function, i. e., a value of the independent variable which when substituted in the function causes the latter to vanish. At least the first few roots of $J_0(x)$ are needed for use with the equations that follow; the first four, represented by $\mu_1$, $\mu_2$, $\mu_3$ and $\mu_4$, respectively, will therefore be given. In the order mentioned they are 2.405, 5.520, 8.654 and 11.792. Fuller tables will be found in the special works on the subject (see, for example, BYERLY, p. 286). Two very important additional formulae related to those found in the standard textbooks but which cannot be derived here are the following

$$\int_0^a xJ_0\left(\mu_m\frac{x}{a}\right)J_0\left(\mu_n\frac{x}{a}\right)dx = 0 \qquad (120)$$

and

$$\int_0^a x\left(J_0\left(\mu_n\frac{x}{a}\right)\right)^2 dx = \frac{a^2}{2}\left[(J_0(\mu_n))^2 + (J_1(\mu_n))^2\right]. \qquad (121)$$

With so much by way of an introduction, we may consider equation (118), following with certain modifications the treatment given by BYERLY (1893). It is desired to find a solution of this equation which fulfills the initial condition $u = f(r)$ when $t = 0$ and the boundary condition for the system in question. We shall here consider as a typical and very useful case the single type of boundary condition, $u = 0$ when $r = a$. As before, when a different partial differential equation was dealt with [(8), p. 32], it will be necessary first to find particular solutions of equation (118) and then to select and combine these particular solutions so as to satisfy the given boundary and initial conditions. In order to find particular solutions we may, as before, assume in a purely tentative manner that a solution exists of the form

$$u = R\,T \tag{122}$$

where $R$ is a function of $r$ alone and $T$ a function of $t$ alone (the product $R\,T$ has nothing to do with the similar product appearing in the gas law). In order to find what forms these functions must have to satisfy the equation, we perform on this tentative value of $u$ the operations indicated in (118) and after rearrangement obtain

$$\frac{1}{D\,T}\frac{\partial T}{\partial t} = \frac{1}{R}\left(\frac{\partial^2 R}{\partial r^2} + \frac{1}{r}\frac{\partial R}{\partial r}\right).$$

By hypothesis, the right-hand side of this equation cannot contain $t$ nor the left-hand side $r$. In order, therefore, that the equation may be true for all values of $t$ and $r$, each side must be equal to a constant. Let this constant be $-\frac{\mu^2}{a^2}$ where $a$ is the radius of the cylinder and $\mu$ is another constant. Two equations result, namely

$$\frac{d\,T}{d\,t} + \frac{\mu^2}{a^2} D\,T = 0 \tag{123a}$$

and

$$\frac{d^2 R}{d\,r^2} + \frac{1}{r}\frac{d\,R}{d\,r} + \frac{\mu^2}{a^2} R = 0. \tag{123b}$$

The first of these equations is readily solved by the routine method employed for linear differential equations of the first order, giving as its solution

$$T = C\,e^{-\frac{\mu^2 D t}{a^2}}$$

where $C$ is an arbitrary constant. The second equation, however, does not yield to such simple treatment. In cases such as this where other methods fail to give a solution, a possible method of procedure is to assume a solution in the form of an infinite series of ascending powers of the independent variable. Before doing so, however, the equation may first be simplified by introducing a new variable $x$ of such a nature that $x = \frac{\mu r}{a}$; the equation then becomes

$$\frac{d^2 R}{d\,x^2} + \frac{1}{x}\frac{d\,R}{d\,x} + R = 0. \tag{124}$$

Now assume that $R$ can be expressed as the sum of a series of terms of the form $a_n x^n$, i. e., by $\Sigma\, a_n x^n$, $n$ being an integer. On this assumption

$$\frac{d^2 R}{d\,x^2} = \sum n\,(n-1)\,a_n\,x^{n-2} \quad \text{and} \quad \frac{1}{x}\frac{d\,R}{d\,x} = \sum n\,a_n\,x^{n-2}.$$

We have, therefore

$$\Sigma\left[n\,(n-1)\,a_n\,x^{n-2} + n\,a_n\,x^{n-2} + a_n\,x^n\right] = 0.$$

In order that this relation may be true for all values of $x$, the coefficient of each power of $x$ must separately be equal to zero. If $k$ be used to represent some particular value of $n$, the term having the exponent $k-2$ will be composed of the first two terms in the last equation, with $k$ substituted for $n$ together with the third one in which the coefficient is $a_{k-2}$. Since the sum of these terms must be zero, i. e., since

$$k(k-1)\,a_k + k\,a_k + a_{k-2} = 0$$

it follows that

$$a_{k-2} = -k^2 a_k.$$

This is the law, therefore, which the coefficients of the series must follow to be a solution of (124.) If $k = 0$, then $a_{k-2}$, $a_{k-4}$, etc. are also all equal to zero, so this value of $k$ may be chosen as the starting point of the series. Let $a_0$ be the coefficient of this term. Then

$$a_2 = -\frac{a_0}{2^2}, \qquad a_4 = \frac{a_0}{2^2 \cdot 4^2} \text{ etc.}$$

and finally

$$R = a_0\left(1 - \frac{x^2}{2^2} + \frac{x^4}{2^2 \cdot 4^2} - \ldots\right).$$

This is not the most general solution of equation (124), since it contains only one arbitrary constant instead of two, but inasmuch as only a particular solution is needed, provided that it can be made to fit the given conditions of the problem, and as this solution turns out to do so, it is unnecessary to proceed farther. Remembering, therefore, the definition of a BESSEL's function of order zero given above, we may write

$$R = a_0 J_0(x) = a_0 J_0\left(\mu \frac{r}{a}\right).$$

Having now solutions of (123a) and (123b) we may combine them as in equation (125) incorporating the two arbitrary constants into a single one, $A$, and we have

$$u = A J_0\left(\mu \frac{r}{a}\right) e^{-\frac{\mu^2 D t}{a^2}} \tag{125}$$

which is a particular solution of (118). But the sum of any number of such terms, having any values of $A$ and $\mu$, will also be a solution.

In the present problem it is necessary in order to satisfy the boundary condition that $u$ should equal zero when $r = a$. This condition is easily fulfilled by selecting as the successive values of $\mu$ in the sum of an infinite series of terms of the form of (125) the roots, $\mu_1$, $\mu_2$, etc. which cause the function $J_0(\mu)$ to vanish. We now have a series

$$u = \sum_{n=1}^{n=\infty} A_n J_0\left(\mu_n \frac{r}{a}\right) e^{-\frac{\mu_n^2 D t}{a^2}} \tag{126}$$

which satisfies the original equation and reduces to zero when $r = a$, thereby satisfying the boundary condition as well. If now a series of values of $A_n$ can be selected which make the series of terms equal to $f(x)$ when $t = 0$, then the initial condition will also be satisfied and the solution will be complete.

Such a series can be found by a method entirely analogous to that employed in determining the FOURIER coefficients of a trigonometric series. Assume that a solution of the form

$$f(r) = A_1 J_0\left(\mu_1 \frac{r}{a}\right) + A_2 J_0\left(\mu_2 \frac{r}{a}\right) + \ldots + A_m J_0\left(\mu_m \frac{r}{a}\right) + \\ + A_n J_0\left(\mu_n \frac{r}{a}\right) + \ldots$$

can be found. Multiply both sides of the equation by $r J_0\left(\mu_n \frac{r}{a}\right) dr$ and integrate between zero and $a$. Three forms of integrals will appear, namely

$$\int_0^a r f(r) J_0\left(\mu_n \frac{r}{a}\right) dr, \qquad A_m \int_0^a r J_0\left(\mu_m \frac{r}{a}\right) J_0\left(\mu_n \frac{r}{a}\right) dr$$

and

$$A_n \int_0^a r \left(J_0\left(\mu_n \frac{r}{a}\right)\right)^2 dr.$$

By equation (120) we find that the second of these three types of integrals reduces to zero. Furthermore, by employing equation (121) with the third type and proceeding exactly in the manner already employed with a FOURIER's series we obtain

$$A_n = \frac{2}{a^2 [(J_0(\mu_n))^2 + (J_1(\mu_n))^2]} \int_0^a r f(r) J_0\left(\mu_n \frac{r}{a}\right) dr.$$

But since $\mu_n$ is a root, $J_0(\mu_n)$ is equal to zero and we have

$$A_n = \frac{2}{a^2 (J_1(\mu_n))^2} \int_0^a r f(r) J_0\left(\mu_n \frac{r}{a}\right) dr. \tag{127}$$

In the case of most importance in physiology the initial distribution of material in the cylinder is $f(r) = u_0$ when $t = 0$. Substituting this value of $f(r)$ in equation (121) and using (119b) we have

$$A_n = \frac{2 u_0}{\mu_n J_1(\mu_n)}.$$

By substituting coefficients of this type in (126), the resulting series, which already satisfies the diffusion equation and the boundary condition, now satisfies the initial condition as well. The complete solution is, therefore

$$u = 2 u_0 \sum_{n=1}^{n=\infty} \frac{J_0\left(\mu_n \frac{r}{a}\right)}{\mu_n J_1(\mu_n)} e^{-\frac{\mu_n^2 D t}{a^2}}. \tag{128}$$

For the case where diffusion is inward from an external concentration of $c$ into an originally empty cylinder the equation assumes a slightly different form, which for brevity may be included in the more general case in which the external concentration is $c$ and the initial internal concentration is $u_0$. The equation applicable to this case, obtained by methods similar to those used on p. 50, may be written (see WILLIAMSON and ADAMS 1919, p. 111 and ANDREWS and JOHNSTON 1924, p. 647)

$$\frac{u-c}{u_0-c}=2\sum_{n=1}^{n=\infty}\frac{J_0\left(\mu_n\frac{r}{a}\right)}{\mu_n J_1(\mu_n)}e^{-\frac{\mu_n^2 D t}{a^2}}. \tag{129}$$

In order to use this equation it is necessary to have not only the values of $\mu_1$, $\mu_2$, etc., some of which were given on p. 115, but those of $J_1(\mu_n)$. The first four of the latter values are 0.5191, —0.3403, 0.2714 and —0.2325, respectively; others may be found from the published tables.

Other equations of importance in dealing with diffusion problems in cylinders may readily be derived from those given. For example, the maximum concentration at any position in the cylinder for cases of outward diffusion, and the minimum concentration for cases of inward diffusion will obviously be found at the axis of the cylinder where $(r=0)$. Remembering that $J_0(0)=1$, it follows immediately from equation (129) that for this particular value of $r$

$$\frac{u-c}{u_0-c}=2\sum_{n=1}^{n=\infty}\frac{1}{\mu_n J_1(\mu_n)}e^{-\frac{\mu_n^2 D t}{a^2}}. \tag{130}$$

WILLIAMSON and ADAMS (1919, p. 114) have calculated by means of this equation a short but useful table of values of $u$ for $r=0$ corresponding to different values of $\frac{Dt}{a^2}$.

To find the quantity of material, $Q$, contained at any time in a cylinder in which diffusion is occurring, it is only necessary to write equation (129) in a form that is explicit for $u$, multiply by $2\pi r L\,dr$ and integrate between 0 and $a$. Since the successive steps in this integration are fully given by ANDREWS and JOHNSTON (1924, p. 649—650) and since no new principles other than an obvious application of equation (119b) are involved, it will be sufficient to give here only the end result for the case in which $u_0=0$, and the length is $L$

$$Q_{0,a}=\pi a^2 L c\left(1-4\sum_{n=1}^{n=\infty}\frac{1}{(\mu_n)^2}e^{-\frac{\mu_n^2 D t}{a^2}}\right). \tag{131}$$

From this equation the average concentration, $\bar{u}$, is obtained by dividing by the volume of the cylinder, $\pi a^2 L$, and the degree of saturation by dividing by the value of $Q$ at equilibrium, namely, $\pi a^2 Lc$.

The degrees of saturation for different values of $D$, $t$ and $a$ have been calculated by HILL (1928, p. 71) and have also been plotted as a useful curve (l. c. p. 70). From these results HILL draws several interesting conclusions. For example, taking $D$ for oxygen in nerve as $4.5 \times 10^{-4}$ and the radius of the frog's sciatic nerve as 0.35 mm. he shows that saturation would be 50 per cent complete in about 10 seconds and 90 per cent complete in 54 seconds. Conversely, a nerve previously saturated with oxygen and then exposed to nitrogen would lose 99 per cent of its contained oxygen in 2.2 minutes. Since the amount of oxygen the nerve can hold at saturation at $20^0$ C is about 0.006 c.c. per c.c., and since its metabolism at rest is of the order of 0.0005 c.c. per c.c. per minute, the amount that would remain at the end of 2.2 minutes, even neglecting metabolism before this time, would suffice for its needs for only a few seconds longer. It follows therefore that when a nerve continues to function in pure nitrogen for 2 hours it cannot be supposed to do so at the expense of a supply of dissolved molecular oxygen.

HILL's calculated curve for diffusion in a cylinder has recently been used by COLLANDER and BÄRLUND (1933) to determine whether the rate of entrance of dissolved substances into the cylindrical cells of *Chara* is of the nature that would be expected if diffusion within the cell were the limiting factor, or whether it is of the sort that would result if the chief obstacle to diffusion were found in a limiting cell membrane. The conclusion is drawn from a comparison of the observed with the two theoretical types of penetration curves that a delay in passing through the cell surface must be the more important of the two factors.

An equation is readily derived for a steady state of symmetrical diffusion in a cylinder. For this case, in which $\frac{\partial u}{\partial t} = 0$, we have from equation (118)

$$\frac{d^2 u}{d r^2} + \frac{1}{r}\frac{du}{dr} = 0.$$

The solution of this equation is

$$u = C_1 \ln r + C_2.$$

As boundary conditions we have the information that when $r = r_1$, $u = u_1$, and when $r = r_2$, $u = u_2$. From this information the constants are readily evaluated and we have

$$u = \frac{(u_1 - u_2)\ln r}{\ln r_1 - \ln r_2} + \frac{u_1 \ln r_2 - u_2 \ln r_1}{\ln r_2 - \ln r_1}. \tag{132}$$

To obtain the amount of material that would flow outward in unit time across any cylindrical surface of unit length within the region to which the equation is applicable, it is only necessary to obtain from (132) the value of $-\frac{\partial u}{\partial r}$ and to multiply it by the diffusion coefficient and the area. The result is

$$Q = \frac{2\pi D (u_1 - u_2)}{\ln r_2 - \ln r_1}. \tag{133}$$

As should be the case for a steady state, this result is independent of $r$.

## 13. Diffusion in spheres.

Cases of diffusion in spheres are not infrequent in physiology, the commonest examples perhaps being diffusion into or out of single spherical cells such as ova, cocci, etc. Cases of this sort evidently involve three dimensions, unless the sphere be so large or the depth to which diffusion extends be so small that the surface may with little error be treated as if it were a plane (the cooling of the earth's crust, for example, may be dealt with in this way). Though on superficial consideration it might be expected that the treatment of diffusion in three dimensions in the case of the sphere would be mathematically more complicated than that in two dimensions in the case of a cylinder, the reverse is true when, as very frequently happens, the diffusion process is symmetrical about the center of the sphere. In this case equations very similar to those already derived for certain one-dimensional processes may be employed.

The general equation, expressed in spherical coordinates, for diffusion in three dimensions has already been given [(10) p. 23]. When diffusion occurs symmetrically about the center of a sphere, $u$ depends only on $r$ and not on $\theta$ and $\varphi$ equation [10] reduces to the simpler form

$$\frac{\partial u}{\partial t} = \frac{D}{r^2}\left(\frac{\partial}{\partial r} r^2 \frac{\partial u}{\partial r}\right). \tag{134}$$

After performing the differentiation indicated within the parenthesis, this equation may be written

$$r\frac{\partial u}{\partial t} = D\left(r\frac{\partial_2 u}{\partial r^2} + 2\frac{\partial u}{\partial r}\right). \tag{134 a}$$

If now a new variable, $v = ru$ be chosen, it will be seen that since $\frac{\partial v}{\partial t} = r\frac{\partial u}{\partial t}$ $\left(\frac{\partial r}{\partial t}\text{ being equal to zero}\right)$ and since $\frac{\partial^2 v}{\partial r^2} = r\frac{\partial^2 u}{\partial r^2} + 2\frac{\partial u}{\partial r}$ equation (134) may be written

$$\frac{\partial v}{\partial t} = D\frac{\partial^2 v}{\partial r^2}. \tag{135}$$

This, however, is an equation of exactly the same form as (8) and can be solved by the same methods. It is only necessary before attempting to do so to be sure that the initial and boundary conditions have been properly expressed in terms of the new variable, $v$.

The solution of equation (135) may first be obtained for a somewhat general case, from which the treatment of several special cases of physiological importance may readily be deduced. Suppose that in a spherical cell of radius $R$ a diffusible substance be originally distributed in some symmetrical fashion about the center in the manner described by $u = f(r)$. Let the concentration of the same substance at the cell boundary, $r = R$, be constantly maintained at the value $c$; this may or may not be the concentration in the external medium, but for present purposes, nothing but the mathematical boundary condition need be considered. In such a system, for all finite values of $u$,

$v = ru = 0$ when $r = 0$; we have therefore to find a solution of equation (135) which will fulfill the conditions

$$\begin{aligned} v &= 0 &&\text{when } r = 0 \\ v &= cR &&\text{when } r = R \\ v &= rf(r) &&\text{when } t = 0 \end{aligned}$$

Now a solution can readily be found for the diffusion equation (8) for the conditions

$$\begin{aligned} u &= 0 &&\text{when } x = 0 \\ u &= c &&\text{when } x = H \\ u &= f(x) &&\text{when } t = 0\,. \end{aligned}$$

by combining equations (23) and (51) after substituting in the latter $c_1 = 0$ and $c_2 = c$. This solution may then be adapted to the present case by an appropriate change of symbols giving

$$\left.\begin{aligned} ur = {} & \frac{2}{R}\sum_{n=1}^{n=\infty}\left(e^{-\frac{n^2\pi^2 Dt}{R^2}}\sin\frac{n\pi r}{R}\int_0^R \lambda f(\lambda)\sin\frac{n\pi\lambda}{R}\,d\lambda\right) \\ & + c\left(r + \frac{2R}{\pi}\sum_{n=1}^{n=\infty}\left(\frac{(-1)^n}{n}e^{-\frac{n^2\pi^2 Dt}{R^2}}\sin\frac{n\pi r}{R}\right)\right). \end{aligned}\right\} \tag{136}$$

Equation (136) gives the concentration, $u$, for any values of $r$, $t$ and $D$, for any symmetrical initial distribution of the solute about the center of the cell which can be represented by $u = f(r)$. The case of most practical importance is that in which an initial concentration of $u_0$ originally exists everywhere within the cell and a concentration of $c$ is constantly maintained at the cell surface. For this case we obtain

$$u = c + \frac{2R}{\pi r}(u_0 - c)\left(\sin\frac{\pi r}{R}e^{-\frac{\pi^2 Dt}{R^2}} - \frac{1}{2}\sin\frac{2\pi r}{R}e^{-\frac{4\pi^2 Dt}{R^2}} + \ldots\right). \tag{137}$$

When $t = \infty$ equation (137) reduces to $u = c$ as it should. Also, since by the ordinary FOURIER expansion within the range in which we are interested

$$r = \frac{2R}{\pi}\left(\sin\frac{\pi r}{R} - \frac{1}{2}\sin\frac{2\pi r}{R} + \ldots\right)$$

when $t = 0$ we have $u = u_0$ as we should.

For many purposes it is convenient to write equation (137) in the form

$$\frac{u - c}{u_0 - c} = \frac{2}{\pi}\sum_{n=1}^{n=\infty}\left(\frac{\frac{R}{r}\sin\frac{n\pi r}{R}}{n(-1)^{n+1}}e^{-\frac{n^2\pi^2 Dt}{R^2}}\right). \tag{138}$$

Except for a slight change in the symbols, this is identical with the equation given by WILLIAMSON and ADAMS (1919, p. 111) for the temperature distribution in a sphere. From it may readily be derived equations to cover the two cases of greatest practical importance in physiology, namely, those in which with an initial uniform concentration of $u_0$ the surface concentration has the constant value zero and in which with an initial uniform concentration of zero the surface concentration has the constant value $c$. The former equation may be applied to cases of escape of substances from and the latter to cases of entrance of substances into a homogeneous spherical cell. A useful numerical table covering the general case, represented by equation (138), is given by WILLIAMSON and ADAMS (l. c., p. 111).

Of all values of $r$ for which $u$ may be calculated, perhaps the most interesting is $r = 0$, since the concentration at the exact center is either the maximum or the minimum existing within the sphere at any given time, according to whether diffusion is proceeding outwards or inwards. On substituting $r = 0$ in equation (138) the indeterminate form $\infty \cdot 0$ results, but by noting that as $r$ approaches 0, $\sin \frac{n\pi r}{R}$ approaches $\frac{n\pi r}{R}$, we obtain a finite limiting value of $\frac{R}{r} \sin \frac{n\pi r}{R}$ which gives for the concentration at the center of the sphere

$$\frac{u - c}{u_0 - c} = 2 \sum_{n=1}^{n=\infty} \left( (-1)^{n+1} e^{-\frac{n^2 \pi^2 D t}{R^2}} \right). \tag{139}$$

This equation has been used by WILLIAMSON and ADAMS (l. c., p. 114) for the preparation of a table which will be found helpful in connection with certain calculations having to do with conditions at the center of a sphere.

It is frequently desirable to be able to calculate the amount of substance that in any given time would cross through a spherical surface at a distance $r$ from the center of the sphere. To do so it is necessary merely to find from equation (137) the value of $\frac{\partial u}{\partial r}$, multiply this value by $-D A d t$ and integrate between 0 and $t$, remembering that for a sphere $A = 4\pi r^2$. In this way we obtain an equation which for the case of most practical importance in which $r = R$, and in which therefore the sine terms are all equal to zero, has the form

$$Q_{0,t} = \frac{8 R^3 (u_0 - c)}{\pi} \left( \frac{\pi^2}{6} - \sum_{n=1}^{n-\infty} \frac{1}{n^2} e^{-\frac{n^2 \pi^2 D t}{R^2}} \right) \tag{140}$$

where for compactness advantage is taken of the relation

$$\frac{\pi^2}{6} = 1 + \frac{1}{4} + \frac{1}{9} + \dots$$

Equation (140), as it should, reduces to zero for $t = 0$ and to $\frac{4}{3} \pi R^3 (u_0 - c)$ for $t = \infty$.

To calculate the amount of solute that at any time $t$ would be enclosed within a spherical surface of radius, $r$, equation (137) is multiplied by $4\pi r^2 dr$ and integrated between 0 and $r$. For the most useful case, which alone need be mentioned, where $r = R$, i. e., where the entire sphere is involved

$$Q_{0,R} = \frac{4}{3}\pi R^3 c + \frac{8 R^3 (u_0 - c)}{\pi}\left(e^{-\frac{\pi^2 D t}{R^2}} + \frac{1}{4} e^{-\frac{4\pi^2 D t}{R^2}} + \ldots\right). \quad (141)$$

This equation, as it should, reduces to $Q_{0,R} = \frac{4}{3}\pi R^3 u_0$ when $t = 0$, and to $\frac{4}{3}\pi R^3 c$ when $t = \infty$. From equation (141) the average concentration within the sphere at any time $t$ may readily be found by dividing it by the volume of the sphere. A useful form of the resulting equation is

$$\frac{\bar{u} - c}{u_0 - c} = \frac{6}{\pi^2}\left(e^{-\frac{\pi^2 D t}{R^2}} + \frac{1}{4} e^{-\frac{4\pi^2 D t}{R^2}} + \ldots\right) \quad (142)$$

which may be compared with the similar equation for a flat sheet given on p. 51. Finally, for the important special case where diffusion is inward from a constant concentration of $c$ into an originally empty sphere, the degree of saturation is given by the equation

$$\frac{\bar{u}}{c} = 1 - \frac{6}{\pi^2}\left(e^{-\frac{\pi^2 D t}{R^2}} + \frac{1}{4} e^{-\frac{4\pi^2 D t}{R^2}} + \ldots\right). \quad (143)$$

The equation for a steady state of symmetrical diffusion in the sphere is sometimes useful and may very readily be derived. For this case, equation (134 a) assumes the form

$$\frac{d^2 u}{d r^2} + \frac{2}{r}\frac{d u}{d r} = 0.$$

A differential equation of exactly this type has already been solved (see p. 69). Using the same solution, but introducing the boundary conditions appropriate to the present case, namely, $u = u_1$ when $r = r_1$ and $u = u_2$ when $r = r_2$ the two constants of integration can be evaluated giving

$$u = \frac{u_2 r_2 - u_1 r_1}{r_2 - r_1} + \frac{r_1 r_2 (u_1 - u_2)}{r (r_2 - r_1)}. \quad (144)$$

To find the amount of material which crosses any spherical surface of radius $r$ in unit time it is only necessary to find $-DA\frac{\partial u}{\partial r}$. Remembering that $A = 4\pi r^2$ we obtain in this way the very simple equation

$$Q_{0,t} = \frac{4\pi D (u_1 - u_2) r_1 r_2}{r_2 - r_1}. \quad (145)$$

## 14. Diffusion subject to external forces.

### a) Diffusion in a gravitational field.

The systems so far considered have been supposed to be governed solely by FICK's law; that is, $u$ has been treated as if it were completely determined by $x$, $t$ and $D$. Strictly speaking, systems of this sort do not exist, since gravitation must alway have some influence on the distribution of freely movable molecules whose masses and volumes are not identical. In ordinary solutions, however, gravitational effects are so slight as to be entirely negligible; they become progressively more conspicuous with increasing size of the particles in colloidal systems; in microscopic suspensions showing Brownian movement they are of predominant importance, while in macroscopic suspensions they almost alone determine the ultimate distribution of the particles. Historically, studies with the microscope and the ultramicroscope of the behavior of particles under the combined influence of diffusive and gravitational forces have played a most important part in the development of modern ideas of the nature of molecular movements and diffusion processes PERRIN 1909, SVEDBERG 1912, WESTGREN 1914, etc.).

While the general case of diffusion in a gravitational field has been treated mathematically by a number of workers (DES COUDRES 1895, MASON and WEAVER 1924, WEAVER 1926, 1927, and FÜRTH 1917, 1927a, b) it is too complex to be considered here except with respect to the final state assumed by such a system when the effects of gravitation and diffusion are exactly balanced. This case is analogous to that of the distribution of gases in the earth's atmosphere, and can be treated by elementary methods. It will be shown below (p. 134) that the same treatment applies equally well to the physiologically important case of diffusion against a convection current.

Imagine a system in which diffusion of particles heavier than water is occurring from below upwards in a column of infinite height in opposition to the force of gravitation. In any horizontal element of unit area and of thickness $dx$ we may consider the number of particles tending to enter and to leave the element under the influence of diffusion on the one hand and of gravitation in the other, and may set up a differential equation in exactly the same manner as that discussed on p. 22 except that concentrations may be measured in terms of the number of particles, rather than of grams or of mols, in unit volume; distance will here be measured from the bottom of the system.

By FICK's law, in time $dt$, the number of particles entering the chosen element of volume from below by diffusion will be $-D\frac{\partial u}{\partial x}dt$, while the number leaving it above will be $-D\left(\frac{\partial u}{\partial x}+\frac{\partial^2 u}{\partial x^2}dx\right)dt$. From the number entering it from below by diffusion must, however, be subtracted the number leaving it under the influence of gravitation, which will be equal to $\gamma u dt$ where $u$ is the concentration at the boundary and $\gamma$ the velocity of fall of a single particle. Similarly from the number of particles leaving the upper surface of the element by diffusion must be subtracted the number entering it under the influence of gravitation, which is $\gamma\left(u+\frac{\partial u}{\partial x}dx\right)dt$. The rate at which the number of particles in the element is increasing is therefore $\left(D\frac{\partial^2 u}{\partial x^2}+\gamma\frac{\partial u}{\partial x}\right)dx$. Otherwise expressed, this is equal to $\frac{\partial u}{\partial t}dx$. We have, therefore, as a description of the process in mathematical form

$$\frac{\partial u}{\partial t}=D\frac{\partial^2 u}{\partial x^2}+\gamma\frac{\partial u}{\partial x}. \tag{146}$$

The velocity of fall $\gamma$ may be found from the relation $\gamma=\frac{\varphi(\varrho-\varrho')g}{f}$ where $\varphi$ is the volume of each particle, $\varrho$ its density, $\varrho'$ the density of water, $g$ the gravitational constant, and $f$ the frictional resistance to the movement of a particle at unit velocity.

While a general solution of equation (146) presents considerable difficulties (see references above) it is very simple after a stationary state has been established, for which $\frac{\partial u}{\partial x}=0$. The equation under these conditions becomes

$$\frac{\partial^2 u}{\partial x^2}+\frac{\gamma}{D}\frac{\partial u}{\partial x}=0$$

which can readily be solved by introducing the new variable $p=\frac{\partial u}{\partial x}$; the resulting linear differential equation of the first order has the solution

$$p=C_1 e^{-\frac{\gamma x}{D}}$$

and this in turn leads to the general solution with two arbitrary constants

$$u=C_1 e^{-\frac{\gamma x}{D}}+C_2. \tag{147 a}$$

The constant $C_2$ is seen to be equal to zero since, for an infinitely high column $u=0$ when $x=\infty$. The constant $C_1$ is found by noting that the total number of diffusible particles in the system, which may be represented by $N$, is equal to $\int_0^\infty u\,dx$. From this relation it follows that $C_1=\frac{N\gamma}{D}$ and the final equation is

$$u=\frac{N\gamma}{D}e^{-\frac{\gamma x}{D}}. \tag{147 b}$$

Regardless of the value of $C_1$, however, it immediately follows that for any two levels

$$\frac{u_2}{u_1}=e^{-\frac{\gamma}{D}(x_2-x_1)}. \tag{148}$$

That is to say, if the natural logarithms of the concentrations observed with the microscope, or otherwise, at different levels be plotted against the distances of these levels from the bottom of the vessel, a straight line, of slope, $-\gamma/D$, should be obtained.

This result was, in fact, obtained by PERRIN (1909) and was one of the earliest proofs that the laws of diffusion apply to visible particles showing Brownian movement. In one series of observations with gamboge particles having a radius of 0.212 $\mu$, observations of concentrations were made at four distances from the bottom of a vessel. The concentrations found, based on a count of 13,000 particles, showed the relative values indicated in the second column. The common logarithms of these numbers, shown in the third column, are seen to have a roughly constant difference; taking the average value of this difference, the theoretical concentrations for each level given in the fourth column may be calculated. The agreement is very good.

| Height in $\mu$ | Relative Concentration (observed) | Logarithm | Relative Concentration (theoretical) |
|---|---|---|---|
| 5 | 100 | 2.0000 | 100 |
| 35 | 47 | 1.6721 | 49 |
| 65 | 22.6 | 1.3541 | 24 |
| 95 | 12 | 1.0792 | 12 |

Similar results were obtained with mastic particles 0.52 $\mu$ in diameter. Though these results have become so familiar as to seem almost self-evident, it was considered most surprising at the time they were first announced that the law that governs the density of the earth's atmosphere should also apply so perfectly to particles which according to PERRIN have a mass $10^9$ times that of the molecules found in the air.

Very similar, but much more extensive observations were later made by WESTGREN (1914) on the ultramicroscopic particles in gold and selenium sols. Certain other results, reported in the same paper, on the rate of displacement of these particles have been mentioned above (p. 91), and must not be confused with the present ones, which have to do with sedimentation equilibrium rather than with a rate of diffusion. In all of WESTGREN's experiments a very good linear relation was obtained between height and the logarithm of the concentration, as may be seen by consulting his numerous tables and figures.

Further important use may be made of experiments of the sort just described. Expressing equation (148) in logarithmic form, using the value of $\gamma$ given above after substituting $\frac{4}{3}\pi r^3$ for $\varphi$, and replacing $D$ by $RT/Nf$ the following equation is obtained

$$\ln \frac{u_1}{u_2} = \frac{4 N \pi r^3 (\varrho - \varrho') g (x_2 - x_1)}{3 R T}. \tag{149}$$

From this equation, if $r$, the radius of the particles can be measured, AVOGADRO's number, $N$, may be calculated. A severe test of the full quantitative applicability of the laws of diffusion to suspended particles is furnished by a comparison of the value of $N$ so calculated with that obtained by other independent methods. PERRIN, who first carried out this test with various particles in suspension, considered his most accurate value to be $6.82 \times 10^{23}$. This value, though of the right order of magnitude is somewhat high, but WESTGREN (l. c.) by the same method obtained a value of $(6.06 \pm 0.20) \times 10^{23}$ which happens to be in perfect agreement with the most generally accepted value, obtained by another method.

In concluding this section it should be pointed out that while equation (148) is theoretically valid and has been shown to hold very satisfactorily in the cases mentioned, it does not necessarily apply in other cases in which complicating factors of various sorts may be present. Thus, BURTON and BISHOP (1922), BURTON and CURRIE (1924) and PORTER and HEDGES (1922) found that in relatively large volumes of several different colloidal suspensions a uniformity of distribution in the main body of the suspension persists unchanged over long periods of time, though PERRIN's type of distribution may be found near the surface. These results are discussed, from a mathematical point of view, by WEAVER (1926, 1927) and by FÜRTH (1927a, b).

### b) Diffusion in a centrifugal field.

The force of gravitation is not sufficiently great to influence to an appreciable extent the diffusion of single molecules, even those of relatively high molecular weight. Its effects become conspicuous only in dealing with larger particles, such as those studied by PERRIN and WESTGREN, each of which is composed of an enormous number of molecules. By the use of centrifugal force, however, which can if desired be made to exceed that of gravitation by hundreds of thousands of times, it becomes possible to employ with molecules, especially with large ones such as those of the proteins, essentially the same methods as those described in the preceding section. These methods in the hands of SVEDBERG and his collaborators (for a recent bibliography see SVEDBERG 1934a, b) have thrown most important light on the molecular weights and other properties of a large number of compounds of physiological significance, and are therefore of particular interest to biologists. While it will be necessary to refer the reader to the numerous original papers of SVEDBERG and his associates for a detailed description both of the methods employed and of the results obtained by their use, a brief discussion of the part played by diffusion processes in several of these methods may be included here.

Apart from the enormously greater complexity of the apparatus required for studies of diffusion in centrifugal fields, there is an important mathematical difference between this case and that involving gravitation alone, namely, that centrifugal force varies with the distance from the axis of rotation, while gravitation is, practically speaking, constant. In spite of this complication, however, it is easy by methods analogous to those discussed in the preceding section to obtain equations for finding molecular weights and molecular radii from the final distribution of a solute of high molecular weight when centrifugal force is balanced against the diffusive tendency of the solute molecules.

Following SVEDBERG, but with the retention of the symbols so far used in the present paper, we may first consider the amount of solute, $dQ$, thrown by centrifugal force towards the periphery of the containing vessel in the time $dt$, and then the amount that in the same time tends to diffuse in the opposite direction in accordance with FICK's law. When a balance has been reached between the two types of movement, they may be equated to each other. It is evident that the centrifugal force acting on one mol of substance of molecular weight $M$ is equal to $x\omega^2 A$ where $x$ is the distance from the axis of rotation, $\omega$ the angular velocity of the centrifuge and $A$ the effective mass of the substance, i. e., the true mass less that of the water displaced by it. $A$ is evidently equal to $M(1 - V\varrho)$ where $V$ is the partial specific volume of the solute and $\varrho$ the density of the solvent. Furthermore, the rate at which material is driven from unit concentration across a given plane is directly proportional to the force applied and inversely proportional to the frictional resistance encountered. For one mol, this frictional resistance may be represented by $F$, and therefore for a concentration of $u$ we have for the amount of material moved across unit area by centrifugal force

$$dQ = \frac{u\,\omega^2\,x\,M\,(1 - V\varrho)\,dt}{F}. \tag{150}$$

Similarly, for the amount moved by diffusion in the opposite direction across unit area, we have by FICK's law [representing $D$ as in equation (3) and remembering that $Nf = F$]

$$dQ = \frac{RT}{F}\frac{du}{dx}dt.$$

Equating the two values of $dQ$, we obtain for an equilibrium between the two forces

$$\frac{du}{u} = \frac{M\,(1 - V\varrho)\,\omega^2\,x\,dx}{RT}$$

and after integration and transposition

$$M = \frac{2\,RT\ln(u_2/u_1)}{\omega^2\,(1 - V\varrho)\,(x_2^2 - x_1^2)}. \tag{151}$$

To determine the molecular weight of a substance to which this method is applicable it is only necessary after equilibrium has been established to measure the relation between the concentrations $u_1$ and $u_2$, of the solution at two levels at the distance $x_1$ and $x_2$ from the axis of rotation, and to know the temperature, the speed of the centrifuge, the partial specific volume of the solute and the density of the solvent. The ingenious methods by which concentrations may be measured in a rotating centrifuge are fully described in the original papers.

As an example of the application of this method to a substance of great physiological importance, the following typical figures obtained by SVEDBERG and FÅHRAEUS (1926) in an experiment with hemoglobin may be cited. Because of its relatively great stability, carbon monoxide hemoglobin was used; its partial specific volume was found by independent measurements to be 0.749. In the experiment in question the speed of the centrifuge was 8708 r.p.m. and $\omega$ was therefore $290.3\,\pi$ per second. The absolute temperature was $293.3^0$. After sedimentation equilibrium had been established (39 hours), measurements of concentration by a photographic method showed in the experiment in question the following relations, selected at random from the large number given in the original paper, between two distances from the axis of rotation and the corresponding concentrations:

| $x$ | $u$ |
|---|---|
| 4.56 | 1.061 |
| 4.51 | 0.930 |

Substituting these values, together with the appropriate value of $R$ in equation (151), the molecular weight of the hemoglobin proves to be 67,670—a figure in excellent agreement with that obtained by other recent workers by entirely different methods (see, for example, on p. 13 a reference to the work of NORTHROP and ANSON).

A somewhat different method, also employed by SVEDBERG, depends not on sedimentation-equilibrium but on sedimentation-rate. Since questions of diffusion are likewise involved in it, a brief discussion of its general principles may be given. It may first be assumed that diffusion is slow in comparison with centrifugalization and that the particles or molecules are all of the same size. Under these conditions a sharp boundary will be maintained between the region occupied by the particles and the pure solvent, and the position and rate of movement of this boundary can be accurately followed. The theoretical rate of movement of the boundary is evidently given by

$$\frac{d\,x}{d\,t} = \frac{\omega^2 x\,M\,(1 - V\varrho)}{F}. \tag{152}$$

On integration and transposition and substitution of $RT/D$ for $F$ the following equation is obtained (SVEDBERG 1925)

$$M = \frac{R\,T \ln (x_2/x_1)}{D\,\omega^2\,(1 - V\varrho)\,(t_2 - t_1)}. \tag{153}$$

It will be noted that although diffusion, as such, has purposely been neglected in the derivation of this equation, the diffusion coefficient nevertheless appears in it; this is merely because it happens to be a convenient, experimentally determinable, measure of frictional resistance. In the more recent papers by SVEDBERG and his associates the simpler relation expressed in equation (152) is used without integration, since over short distances the velocity of sedimentation, $\frac{dx}{dt}$

can be measured directly with little error. Transposing equation (152) and introducing the value of $F$ already used we obtain

$$M = \frac{R T s}{D (1 - V \varrho)} \tag{154}$$

where $s$, the so-called specific sedimentation velocity, which is equal to $\frac{dx}{dt} / \omega^2 x$, is a characteristic constant for every molecular species at a given temperature and for a given solvent. For comparative purposes it is frequently advantageous to calculate it for water at 20° C by means of the relation

$$S_{20} = \frac{dx}{dt} \cdot \frac{1}{\omega^2 x} \cdot \frac{\eta}{\eta_0} \cdot \frac{1 - V \varrho_0}{1 - V \varrho}$$

where $\eta_0$ and $\varrho_0$ are the viscosity and density respectively of water at 20° C. $S_{20}$ is evidently the velocity of sedimentation of a molecule or a particle in water at 20° C in a centrifugal field of unit strength.

It will be noted that if $F$, the molar frictional resistance be substituted for $RT/D$ in equation (154) the following relation is obtained

$$F = \frac{M (1 - V \varrho)}{S}. \tag{155}$$

But if a large molecule be spherical, then the value of $F$ calculated from Stokes' law, namely

$$F = 6 \pi \eta N \left( \frac{3 M V}{4 \pi N} \right)^{\frac{1}{3}} \tag{156}$$

should agree with that obtained from the specific sedimentation velocity; otherwise there will be a discrepancy. By taking the ratio of the two values, a measure is available of the extent to which a given molecule departs in form from a sphere. For Bence-Jones protein, Svedberg and Sjögren (1929) found an almost perfect agreement of the two values, namely, $2.48 \times 10^{16}$ and $2.49 \times 10^{16}$ respectively. For egg albumin the ratio obtained was 1.06, which likewise indicates a spherical shape for this molecule.

Where the diffusion of the solute is not very slow as compared with the displacement of the boundary, the latter will become blurred during the course of an experiment, the more so the longer the experiment is continued. Under certain conditions, however, equation (153) can still be applied to such a system. If the sedimentation is rapid enough to permit a clear zone of solvent to be formed at the inner end of the column of liquid, so that reflection of particles from this end can be assumed to be absent, and if the cell be deep enough so that reflection of particles from the other end can also be neglected during a short experiment, then the system may be treated as if it were infinite in extent, and equation (95) can be applied to the diffusion process which gradually blurs the boundary. By this equation it has already been shown (p. 95) that in systems to which it applies the concentration at the position of the original boundary between solution and solvent must remain constant at $u_0/2$. The position of the boundary that would have been visible had no diffusion taken place can therefore be found by determining from photographs, by methods described in the original papers, the level at which $u = u_0/2$. Equation (153) can then be applied as before. Furthermore, the progress of blurring with time gives an independent means of determining the diffusion coefficient of the substance in question (SVEDBERG and NICHOLS 1927).

Sometimes the radius rather than the molecular weight of the molecules or particles under investigation is desired. This is readily obtained by using the alternative value of $A$, the effective mass of one mol of solute, namely

$$A = N\varphi(\varrho_P - \varrho)$$

where $\varphi$ is the volume of a single molecule, $N$ the AVOGADRO number, $\varrho_P$ the density of a solute molecule and $\varrho$ that of water. After substituting $\frac{4}{3}\pi r^3$ for $\varphi$ and using the resulting value of $A$ in place of $M(1 - V\varrho)$ in (152), at the same time introducing into the equation the value of $F$ obtained by STOKES' law, namely $F = 6\pi\eta r N$, and finally integrating, the following equation results

$$r = \sqrt{\frac{9\eta \ln(x_2/x_1)}{2(\varrho_P - \varrho)\,\omega^2(t_2 - t_1)}}\,. \tag{157}$$

A third method of determining molecular weights by the centrifugal method depends on the diminution with time of the concentration throughout the entire suspension when it is centrifuged in a sector-shaped vessel, in which both the width of the vessel and the centrifugal force increase in the direction of the periphery. Under these conditions, it is easy to show that the diminution of concentration everywhere occurs at the same rate, and that the relation between the concentration $u_1$ which exists when the boundary between the suspension and the pure solute is at the position $x_1$ and the concentration $u_2$ when it has moved to the new position $x_2$ is

$$u_2 = u_1\left(\frac{x_1}{x_2}\right)^2. \tag{158}$$

The following derivation of this relation is somewhat fuller than that given by SVEDBERG and RINDE (1924).

Equation (152) may, for simplicity, be written

$$\frac{dx}{dt} = kx$$

which when integrated becomes

$$kt = \ln\frac{x}{x_0} \tag{159}$$

where $x_0$ is the position occupied by a given particle when $t = 0$. Consider now two particles, one starting at $x_1$ and another at $x_1 + dx$.

At the end of the time $t$, if the new positions of the two particles be represented by $x_2$ and $x_2'$ respectively, we have by equation (159)

$$\frac{x_2}{x_1} = \frac{x_2'}{x_1 + d\,x} \quad \text{or} \quad x_2' - x_2 = \frac{x_2}{x_1}\,d\,x\,.$$

In other words, the distance of their separation in the direction of $x$ has increased in the proportion $x_2/x_1$. At the same time, the length of the annulus of the sector in which the two particles lie has increased in the same ratio. Since a given number of particles has therefore during the time $t$ been transferred to a volume $(x_2/x_1)^2$ times as great as before, the relation of the two concentrations shown in equation (158) immediately follows. That the decrease of concentration must everywhere be the same is seen by considering any two other particles having the initial positions $a\,x_1$ and $a\,x_1 + d\,x$, respectively, and again applying equation (159). If their new positions be designated by $x_2''$ and $x_2'''$ we have

$$x_2''' - x_2'' = \frac{x_2''}{a\,x_1}\,d\,x$$

but since, by the same equation $\ln\,(x_2''/a\,x_1) = \ln\,(x_2/x_1)$ this relation is identical with that already found.

By combining equations (153) or (157) with (158) new equations may therefore be obtained which permit $M$ and $r$ to be calculated from observations of concentrations rather than of distances of movement. Photographic methods are particularly well adapted to the making of such observations, since in the equations employed only the ratios of two concentrations and not the concentrations themselves are involved, and measurements may therefore be made in any convenient, purely arbitrary, units. By means of this method, SVEDBERG and RINDE (1924) obtained values for the radii of colloidal gold particles which were in satisfactory agreement with those found by other methods. The principle is also of importance for making certain corrections that become necessary when considerable movements are observed in a sector-shaped vessel.

When particles of more than one size are present, or when electrical charges must be taken into account, as in a solution of a protein at a distance from its isoelectric point, the necessary mathematical treatment becomes considerably more complicated. For further details concerning these and many other interesting problems, as well as for the values so far obtained of the molecular weights of substances of physiological importance, etc., the reader is referred to the original papers of SVEDBERG and his collaborators. Of particular interest to biologists is the recent paper by SVEDBERG and HEDENIUS (1934) in which all the known data concerning the sedimentation constants of the respiratory proteins of animals have been grouped according to the zoological classification of the animals in question. Striking evidence is obtained that biological relationship is usually associated with similarity or identity in the sedimentation constants of the respiratory proteins. A further fact, of much physiological significance is that when the respiratory proteins have low molecular weights, they are always found to be enclosed in corpuscles; otherwise they would doubtless soon escape from the blood. Conversely, when the proteins are dissolved in the plasma, they nearly always have very high molecular weights, the giant pigment molecules themselves, in a sense, playing the role of blood corpuscles.

### c) Diffusion against a convection current.

Many cases of diffusion in the bodies of organisms are known to be associated with convection currents. For example, the diffusion of gases in the higher animals from the capillaries to the tissues, or vice-versa, must frequently take place either with or against a simultaneous flow of liquid across the capillary wall. Similar cases arise in connection with osmotic exchanges in single cells; when, for example, a cell is exposed to a hypertonic solution of a penetrating solute, and water and solute at first move in opposite directions and then later, after a definite minimum volume has been reached, in the same direction. The same problem is also involved in the case of the non-living membranes studied by MANEGOLD and SOLF (1932). While a completely general treatment of problems of this sort present the same difficulties as those that arise in connection with diffusion in a gravitational field, and physiological applications of any great complexity seem as yet to be lacking, the behavior of diffusion-convection systems, after a stationary condition has been reached is very simple—indeed, some of the equations already derived for diffusion in gravitational fields may be directly applied to systems of this type. One practical use of such equations in connection with the determination of diffusion coefficients has been suggested by SOMERS (1912), and several additional points of interest have arisen in connection with an application of diffusion-convection systems by HERTZ (1922, 1923) to the separation of mixtures of gases having unequal diffusion rates.

To illustrate the principle underlying the method of HERTZ, suppose that two substances, $A$ and $B$, having different diffusion coefficients are allowed to diffuse simultaneously from the constant concentrations $a$ and $b$, respectively, across a region of thickness $H$, at the farther boundary of which they are continually removed, thus maintaining for each a concentration of zero at the plane $x = H$. At any given level, $x$, the ratio of the concentrations of the two substances, $u_1/u_2$, will in general, vary in a complicated way that depends upon $x$, $t$ and $D$; these ratios may if desired be calculated by means of equation (52), p. 58. For a steady state, however, by equation (13),

$$u_1 = a - \frac{a}{H} x \quad \text{and} \quad u_2 = b - \frac{b}{H} x$$

and it is obvious that neither $D$ nor $t$ now has any influence upon the concentration of either substance, and that the concentration ratio $u_1/u_2$ is everywhere the same as in the original mixture. Furthermore, while the absolute *amounts* of the two substances that will cross unit area of a given plane in unit time are different, being in fact, $\frac{D_1 a}{H}$ and $\frac{D_2 b}{H}$ respectively, the ratio of these amounts is constant and independent of position, and the difference between the two diffusion coefficients therefore provides no possibility for the separation of the substances.

When, on the other hand, diffusion occurs against a convection current, then the ratios both of the concentrations and of the rates of diffusion of the two substances must depend upon the position at which the observation is made. This is apparent from the following mathematical considerations. Starting with exactly the same differential equation as that used for diffusion against the force of gravitation, which fits the present case as well, except that $v$, the known velocity of the convection current may be substituted for $\gamma$, the rate of fall in the former equation, we obtain, as before, the general type of solution represented by equation (147a). In the present case, however, instead of a system of infinite height we are dealing with one of thickness $H$, and we have the boundary condition, $u = 0$ when $x = H$. From this information the constant $C_2$ may be evaluated, giving

$$u = C_1 \left( e^{-\frac{vx}{D}} - e^{-\frac{vH}{D}} \right). \tag{160}$$

Also from the other boundary condition that when $x = 0$, $u = a$ (taking the first substance as typical of the behavior of both)

$$C_1 = \frac{a}{1 - e^{-\frac{vH}{D}}}.$$

In general, for large values of $vH/D$, and even for smaller values when merely the ratio of two such constants is involved, as it is in equation (162) below, $C_1$ may with sufficient accuracy be taken as equal to $a$. Now by Fick's law, the amount of substance $Q$ that will in unit time cross unit area at the level $x = H$ is obtained by differentiating equation (160) with respect to $x$ and multiplying the concentration gradient so obtained by $-D$, giving

$$Q = v C_1 e^{-\frac{vH}{D}}. \tag{161}$$

For two substances, therefore, diffusing from constant concentrations of $a$ and $b$, and having diffusion coefficients, $D_1$ and $D_2$, respectively we obtain from (161)

$$\frac{Q_1}{Q_2} = \frac{a}{b} e^{-vH\left(\frac{1}{D_1} - \frac{1}{D_2}\right)}. \tag{162}$$

It follows that in the presence of a convection current the relative quantities of the two substances that leave the system must depend on the values of $H$ and of $v$ chosen for the experiment, as well as upon those of the diffusion coefficients. By taking $v$ or $H$, or both, sufficiently large, the separation of the two substances may theoretically be made as complete as desired. On the other hand, it will be noted from equation (161) that the absolute yield of a given substance will be decreased by increasing either $v$ or $H$, but it will be decreased less by increasing $v$ than $H$. As a compromise, therefore, to secure both a fairly good yield and a fairly good separation, HERTZ recommends taking $v$ large and $H$ small. By means of this interesting type of system HERTZ was able, starting with a mixture of helium and neon containing 30 per cent of helium to separate the latter substance so completely that the spectrum of neon could no longer be detected by rather crude methods. According to HERTZ, the same principle has also applied by F. FISCHER to the separation of diffusible solutes, and the practical possibilities of the method seem as yet by no means to have been exhausted.

## 15. Diffusion of a changing amount of substance.

In the cases so far discussed it has been assumed that the amount of diffusing substance remains constant. It frequently happens, however, in actual physiological diffusion processes, that a diffusing substance is consumed by a cell or a tissue through which it passes; oxygen is the most conspicuous example of a substance of this type. Conversely, a substance such as carbon dioxide may diffuse outward from a cell or tissue in which it is everywhere being produced. While the general problem of the diffusion of a changing amount of substance presents very formidable mathematical difficulties, certain special cases, which fortunately include several of considerable physiological interest, may be solved with ease. Two will be dealt with, namely, that in which the amount of the diffusing substance changes linearly with time, and that in which it changes exponentially.

### a) Rate of change linear.

An example of a diffusion process in which the diffusing substance is consumed at a constant rate in a region through which it passes is furnished by the behavior of oxygen in those cells and tissues in which the rate of respiration, within fairly wide limits, may be assumed to be almost independent of the oxygen tension. This type of respiration is believed to be not uncommon, and although the lack of dependence of rate upon tension is perhaps never entirely complete, and is certainly far from being so at very low tensions, it is nevertheless frequently permissible to make the assumption in question. When it is allowable, the mathematical treatment of the problem is as follows. (It may be noted that if it be known or assumed that all parts of the tissue are respiring, cases of this sort, whether involving flat sheets, cylinders or spheres, may be treated somewhat more simply than by the method here given. This limitation, however, is sometimes undesirable, and will therefore be avoided.)

### α) Diffusion in a flat sheet.

Consider first a flat sheet of tissue of thickness $H$. Suppose that oxygen diffuses into it from both sides from a solution having the constant concentration, $c$. Since the system is symmetrical, each half behaves as if it were separated from the other half by an impermeable partition at the level $x = H/2$ where $\frac{\partial u}{\partial x}$ must evidently be equal to zero. We may therefore take as the boundary conditions $u = c$ when $x = 0$ and $\frac{\partial u}{\partial x} = 0$ when $x = H/2$. Next consider a thin layer lying parallel to the surface of the tissue, of thickness $dx$ and of volume $A dx$, where $A$ is the area of the sheet. The thin layer will consume oxygen at the rate $\alpha A dx$ units of amount per unit of time, if $\alpha$ represent the constant rate of consumption of oxygen of a unit volume. As diffusion progresses, a steady state will tend to be established in which the rate of escape of oxygen from the layer in question must be equal to the rate of entrance less the rate of consumption. In other words

$$-DA\left(\frac{du}{dx} + \frac{d^2u}{dx^2}dx\right) = -DA\frac{du}{dx} - \alpha A dx \tag{163}$$

or

$$\frac{d^2u}{dx^2} = \frac{\alpha}{D}.$$

By successive integrations we obtain

$$\frac{du}{dx} = \frac{\alpha}{D}x + C_1$$

and

$$u = \frac{\alpha}{2D}x^2 + C_1 x + C_2.$$

The constants $C_1$ and $C_2$ in the last two equations are evaluated from the information contained in the two boundary conditions, giving

$$u = c - \frac{\alpha}{2D}(Hx - x^2). \tag{164}$$

From this equation, first derived by WARBURG (1923), the concentration at any depth, $x$, lying between 0 and $H/2$ may readily be found.

The problem may also be stated in the reverse form and an equation may be found to determine the depth, $x$, at which some given concentration $u$ exists. This equation indicates the existence for any value of $u$, of two values of $x$ symmetrically placed with respect to $H/2$. It is

$$x = \frac{H}{2} \pm \sqrt{\frac{H^2}{4} - \frac{2D}{\alpha}(c-u)}. \tag{165}$$

It is of interest to determine how thick a sheet of tissue would have to be in order that with a given diffusion coefficient and a given rate of oxygen consumption an oxygen concentration of zero would just be attained at its mid-plane. Substituting in equation (165) $x = H/2$ and $u = 0$ we obtain

$$H = \sqrt{\frac{8Dc}{\alpha}}. \tag{166}$$

This equation was employed by WARBURG to determine the maximum thickness of a sheet of tissue that could be adequately supplied with oxygen at any given external tension. For such a calculation two pieces of information, in addition to that concerning the external tension are necessary, namely, the appropriate diffusion coefficient of the gas and the rate of oxygen consumption by unit volume of the tissue. In the case of oxygen consumption in slices of liver tissue, the value, given by BARCROFT and SHORE, of $5 \times 10^{-2}$ c.c. of oxygen per c.c. of tissue per minute was used by WARBURG. For the diffusion coefficient (permeability coefficient) it was necessary to employ as a plausible value a figure obtained by KROGH for muscle tissue, namely, $1.4 \times 10^{-5}$ c.c. of oxygen ($0^0$, 760 mm.) per $cm.^2$ per minute with a pressure gradient of 1 atmosphere per cm. Using these figures in equation (166), WARBURG found that with an external atmosphere of pure oxygen the greatest thickness of a sheet of tissue that would permit the penetration of oxygen to its center was 0.47 mm.; with an external partial pressure of oxygen of 0.2 atmosphere the corresponding thickness was found to be 0.21 mm.

These figures were put to a test by MINAMI (1923) as follows. Using slices of liver tissue 0.23 to 0.30 mm. in thickness he first found in a control experiment that the oxygen consumption was the same at partial pressures of this gas of 700 and of 350 mm. respectively, thus showing that at the higher pressure the supply was certainly adequate. He then used sections of increasing thickness and measured the rate of oxygen consumption in $mm.^3$ per mg. of tissue per hour with the following results:

| Thickness of section in mm. | $O_2$ consumed |
|---|---|
| 0.21 | 8.8 |
| 0.30 | 8.8 |
| 0.31 | 9.4 |
| 0.50 | 7.8 |
| 0.95 | 5.8 |
| 1.24 | 5.9 |

It will be noted that the oxygen consumption began to fall off at a thickness somewhere between 0.31 and 0.50 mm. This result is in good agreement with the calculation of WARBURG that for a tension of 760 mm. the maximum thickness of a slice of tissue that can be adequately supplied with oxygen is 0.47 mm.

The same general type of mathematical treatment is possible in the case of the outward diffusion of a substance produced in the tissue. The only difference is that in this case the outward flow from a given layer is equal to the inflow plus the amount produced instead of minus the amount consumed. Setting up an appropriate differential equation and solving it as before the result is

$$u = c + \frac{\beta}{2D}(Hx - x^2) \tag{167}$$

where $\beta$ represents the amount of substance produced per minute by unit volume of tissue and $c$ represents a fixed external concentration of the substance in question; frequently this value is zero. For the maximum concentration within the tissue, i. e., that at the level $x = H/2$,

$$u_{\max} = c + \frac{\beta H^2}{8D}. \tag{168}$$

WARBURG considered the following case: In a section of liver respiring in pure oxygen $H$ was $4.7 \times 10^{-2}$, $D$ for carbon dioxide was taken as 35 times its value for oxygen (KROGH) and $\beta$ for carbon dioxide as $5 \times 10^{-2}$. With an external $CO_2$ tension of zero (i.e., $c = 0$) these figures when introduced into equation (168) give as the maximum carbon dixoide tension in the center of the tissue 0.028, i. e., about 3 per cent of an atmosphere.

### $\beta$) Diffusion in a cylinder.

Very similar problems arise in connection with cells and masses of tissue of other shapes than flat sheets. A most important case is that of the cylinder; this case is involved in studies of the metabolism of nerves and of cylindrical muscles and individual muscle fibers, as well as in connection with diffusion processes from or into the cylindrical capillaries. The mathematical treatment of this problem differs only in details from that already discussed. The chief difference in the two cases is that here the diffusing substance must be considered to enter into and to escape from cylindrical shells of thickness $dr$ and of surface $2\pi Lr$, where $L$ is the length of the cylinder in question. For the case of entrance into a cylinder an equation corresponding to (163) for a flat sheet is similarly derived, namely

$$2DL\pi(r - dr)\left(\frac{du}{dr} - \frac{d^2u}{dr^2}dr\right) = 2DL\pi r\frac{du}{dr} - 2\alpha L\pi r\,dr.$$

(It will be noted that here diffusion is in the direction of decreasing $r$; consequently the minus sign usually present in FICK's equation is changed to plus.) Dropping infinitesimals of higher order than the first, the equation reduces to

$$r\frac{d^2u}{dr^2} + \frac{du}{dr} = \frac{\alpha}{D}r.$$

This equation can be solved by introducing a new variable $p = \frac{du}{dr}$ which converts it into a linear equation of the first order, namely

$$\frac{dp}{dr} + \frac{1}{r}p = \frac{\alpha}{D}$$

from which the solution

$$r p = \frac{\alpha}{2D} r^2 + C_1$$

is obtained, and on substituting the value of $p$ and integrating again

$$u = \frac{\alpha}{4D} r^2 + C_1 \ln r + C_2. \tag{169}$$

The two constants of integration are evaluated from the information given by the two boundary conditions, namely, that $\frac{du}{dr} = 0$ when $r = 0$ and $u = c$ when $r = R$, leading finally to the equation

$$u = c - \frac{\alpha}{4D} (R^2 - r^2). \tag{170}$$

This equation, except for a slight difference in the symbols employed, is the one derived independently at about the same time for use with nerves by FENN (1927) and by GERARD (1927) and for use with muscles by MEYERHOF and SCHULZ (1927). As a single example of the manner in which it may be applied to a physiological problem, the following case dealt with by FENN (l. c.) may be mentioned. It was desired to determine whether under given experimental conditions an adequate supply of oxygen is available in the innermost parts of a nerve. As a maximum value for frog nerve FENN took $R = 0.1$ cm. and as values of $D$ and of $\alpha$, $1.4 \times 10^{-5}$ and $1.23 \times 10^{-3}$, respectively. Introducing these values into equation (170), it appears that with an external oxygen tension of 760 mm. the tension at the center of the nerve under the conditions in question must be 590 mm.; with an external oxygen tension approximately that in ordinary air, on the other hand, the internal tension must just reach zero. The radius at which the minimum internal tension would just reach zero in an atmosphere of pure oxygen may similarly be shown to be 0.213 cm., which is about that of the largest dogfish nerves.

Only a slightly different treatment is needed for an important physiological problem discussed by KROGH, namely, the distance to which oxygen can diffuse from the capillaries in which its tension is $c$, given a certain rate of consumption of oxygen by the surrounding tissues and a certain diffusion coefficient for the gas in these tissues. In this case, a cylindrical shell lying between $r$ and $r + dr$ is considered where $r$ is now greater than the radius of the capillary which may here be designated by $R_1$. If the distance between the centers of two similar capillaries be called $2 R_2$ then the boundary conditions are $\frac{du}{dr} = 0$ when $r = R_2$ and $u = c$ when $r = R_1$. Proceeding as before with these data we again obtain equation (169), but since the boundary conditions are different, the constants of integration are different, and the new solution is somewhat more complicated than previously, namely

$$u = c - \frac{\alpha}{D} \left( \frac{R_2^2}{2} \ln \frac{r}{R_1} - \frac{r^2 - R_1^2}{4} \right). \tag{171}$$

This, except for a difference in the symbols, is the equation used by KROGH (1919 b) to determine the difference in oxygen tension between the capillaries and the regions lying midway between them that would be necessary to prevent a deficiency of oxygen from occurring in any part of the tissue. If the difference in tension between the wall of the capillary and a point at a distance of $R_2$ from its center be represented by $c - u$, then knowing the average distance between two adjacent capillaries ($2\,R_2$), which KROGH obtained by direct measurement, knowing also the radius of the capillary, which he estimated roughly for different species of animals from the diameters of their erythrocytes, and knowing finally the rate of oxygen consumption of the tissues, which he obtained, also very roughly, from that of the organism as a whole, it is possible by taking an appropriate value of $D$ to make any desired calculation. For $D$ KROGH used a value (in the units mentioned above) of $1.64 \times 10^{-5}$ for warm-blooded animals at $37^0$ C, and one of $1.33 \times 10^{-5}$ for cold-blooded animals at $20^0$ C. Introducing the appropriate numerical values into equation (171), the theoretical tension differences between the blood in the capillaries and the deeper parts of the surrounding tissues were obtained. By taking $u = 0$ these same values give the minimum head of oxygen pressure in the capillary that would just supply oxygen to all parts of the tissue, namely:

| Species | Difference in tension in mm. Hg |
|---|---|
| Cod . . . | 0.4 |
| Frog . . | 0.25 |
| Horse . . | 0.1 |
| Dog . . . | 0.2 |
| Guinea-pig | 0.3 |

Making all possible allowances for errors in the admittedly crude methods used for estimating the numerical values used, it is impossible to escape the conclusion drawn by KROGH that the head of oxygen pressure necessary to supply this substance by diffusion to all parts of the muscular tissue, even during severe work, must be extremely low.

### γ) Diffusion in a sphere.

The same type of problem again arises in connection with diffusion in spheres, and the equations that may readily be obtained for bodies of this form have found application in studies on the respiration of spherical ova, bacteria, etc. Mathematically the problem is almost exactly the same as the one last discussed except that the formula $4\,\pi r^2$ for the surface of a sphere replaces that for the surface of a cylinder. Proceeding as before, we obtain for diffusion into a sphere the relation

$$4 D\pi (r - dr)^2 \left(\frac{du}{dr} - \frac{d^2 u}{dr^2} dr\right) = 4 D\pi r^2 \frac{du}{dr} - 4\,\alpha\pi r^2 dr$$

and after simplifying and dropping infinitesimals of higher order

$$\frac{d^2 u}{dr^2} + \frac{2}{r}\frac{du}{dr} = \frac{\alpha}{D}.$$

Introducing the new variable, $p = \frac{du}{dr}$, and employing the integrating factor $r^2$, a solution

$$pr^2 = \frac{\alpha}{3D} r^3 + C_1$$

is obtained, and on integrating again after the substitution of $\frac{du}{dr}$ for $p$

$$u = \frac{\alpha}{6D} r^2 - \frac{C_1}{r} + C_2.$$

Evaluating the constants of integration from the information that $u = c$ when $r = R$ and $\frac{du}{dr} = 0$ when $r = 0$, the equation finally becomes

$$u = c - \frac{\alpha}{6D} (R^2 - r^2) \qquad (172)$$

which differs from the equation for a cylinder only in containing the numerical factor 6 in place of 4. Harvey (1928) took advantage of this similarity in dealing with bacteria of a form intermediate between that of spheres and of rods, and as a rough approximation used the factor 5.

The applicability of equation (172) to the diffusion of oxygen into small spherical cells has been discussed briefly by HARVEY (1928), and by SHOUP (1929) and very fully and with references to the earlier literature by GERARD (1931). In general, it seems that the critical external concentration below which oxygen consumption begins to decrease in small single cells is usually many times greater than that calculated by equation (172) from the known rate of respiration of the cell and any reasonable assumed value for $D$ within the cell—in other words, diffusion in such cases can scarcely be the factor that first limits oxygen consumption. There are various possible, and indeed plausible, explanations of the discrepancy between the observed results and those calculated by the simple diffusion equation; for a full discussion of these and for a mathematical treatment of more complicated cases the reader is referred to GERARD (1931) and to RASHEVSKY (1933).

**b) Rate of change exponential.**

The group of cases just discussed, where a diffusing substance disappears at a constant rate, is easy to deal with mathematically, since it necessarily leads to a steady state, and in consequence to an ordinary differential equation in two variables. Another case of probable future physiological importance, though its practical applications seem so far to have been confined to the field of physics, is the diffusion of a substance which disappears in an exponential manner according to the law governing monomolecular reactions, and represented graphically by the well-known "die away" curve. The case actually dealt with experimentally by physicists is that of the diffusion of decomposing radioactive materials (WALLSTABE 1903, HOFFMAN 1906,

Ramstedt 1919), but many analogous cases must undoubtedly occur in physiology. Though monomolecular reactions in the strictest sense are probably very rare, it is known that bimolecular reactions in which one of the reacting substances is present in large excess (for example, water in the inversion of cane sugar, or in numerous other important hydrolytic processes) follow the same mathematical law, and the diffusion equations would therefore be the same in both cases.

When a diffusing substance disappears exponentially, its total amount never becomes equal to zero, and consequently there is no possibility of the establishment of the steady state which greatly simplified the treatment of diffusion with a constant rate of disappearance. It is easy, however, to deal with the present case by another method, which leads to equations very similar to some already derived for other purposes. To take a concrete problem, dealt with by Ramstedt (1919), suppose that a layer, of depth $h$, of a solution of a radioactive material be placed in a vessel of uniform cross section and that above it be placed a layer of water of equal depth (i.e., $H = 2h$). It is required to find the concentration $u$ in the system for any values of $x$, $t$ and $D$, and for any exponential rate of decomposition of the substance.

In a monomolecular reaction the rate of disappearance of a substance at any instant is proportional to its concentration. Let $\lambda$ be the constant of proportionality. Then if, as before, a differential equation analogous to (8) be set up to represent the rate of increase of concentration in any elementary layer at right angles to the direction of diffusion, and of thickness $dx$, it is necessary merely to subtract $\lambda u$, the rate of change of concentration due to decomposition, from the right-hand side of (8) which represents the rate of increase due to diffusion. The resulting equation is

$$\frac{\partial u}{\partial t} = D\frac{\partial^2 u}{\partial x^2} - \lambda u. \qquad (173)$$

Now introduce a new variable $v$ of such a nature that $u = ve^{-\lambda t}$. Expressing equation (113) in terms of $v$, we have

$$\frac{dv}{dt} = D\frac{d^2 v}{dx^2}. \qquad (174)$$

This equation, however, is nothing but the ordinary diffusion equation in terms of a new variable, $v$. It follows, therefore, that after expressing the initial and boundary conditions in terms of $v$ rather than of $u$ an equation already available may be used for the treatment of this problem. The initial condition is evidently identical for both variables, since when $t = 0$, $u = v$. The same is true for the boundary conditions, since when $\frac{\partial u}{\partial t} = 0$, $\frac{\partial v}{\partial t} = 0$. The solution of the problem is therefore obtained merely by substituting $v$ for $u$ in equation (28). Finally, to obtain a solution terms of $u$ rather than of $v$, the right-hand side of the resulting equation is multiplied by $e^{-\lambda t}$ and the result is

$$u = e^{-\lambda t}\left[\frac{u_0 h}{H} + \frac{2 u_0}{\pi}\sum_{n=1}^{n=\infty}\left(\frac{1}{n}\sin\frac{n\pi h}{H}\cos\frac{n\pi x}{H}e^{-\frac{n^2\pi^2 Dt}{H^2}}\right)\right].$$

In RAMSTEDT's experiments, it was desired to find $D$ from a measurement at the end of a given time of the observed amounts of diffusing substance in the upper and lower halves of the diffusion vessel. By exactly the same method as that previously employed (p. 44) the following equation is obtained in which the plus sign is used for the lower and the minus sign for the upper half of the vessel

$$Q = e^{-\lambda t}\left[\frac{u_0 A H}{4} \pm \frac{2 u_0 A H}{\pi^2}\left(e^{-\frac{\pi^2 Dt}{H^2}} + \frac{1}{9}e^{-\frac{9\pi^2 Dt}{H^2}} + \ldots\right)\right]. \quad (175)$$

If, now, continuing to follow the treatment on p. 45 the ratio of the difference to the sum of the two values of $Q$ be taken, the common factor $e^{-\lambda t}$ disappears and exactly the same equation is obtained as that first used by LOSCHMIDT (1870a, b). By means of this equation RAMSTEDT readily obtained the desired value of the diffusion coefficient of radium emanation in water.

### c) Other cases.

In addition to the two relatively simple cases of the diffusion of a changing amount of substance discussed in the present section, others of a more complex nature, including several of considerable physiological interest, have received mathematical treatment which, however, because of its complexity, cannot be reproduced here. Those who wish to follow this subject farther are referred, in particular, to the valuable paper of HILL (1928), who discusses, among other subjects the following: the simultaneous diffusion of oxygen and lactic acid in opposite directions, the diffusion of oxygen into a region showing an oxygen debt, the diffusion of oxygen into a region in which recovery is everywhere necessary before the oxygen can advance, etc. Another problem of the same sort, of much physiological interest and of considerable mathematical complexity, is that of the simultaneous diffusion and chemical combination with hemoglobin of oxygen in the erythrocyte; this problem has been discussed by ROUGHTON (1932). Still another is the combination of simple diffusion processes with surface conditions of different sorts in cases in which the manner of passage of the diffusing material into regions in which the simple diffusion laws are applicable is not that so far postulated. As examples of this type of problem of actual or potential physiological interest may be mentioned the cases considered by RASHEVSKY (1933) and by McKAY (1930, 1932a, b). Other examples of the same sort could be mentioned, but obviously the proper place to study them is in the original literature where their various complexities are dealt with in a manner that would be impossible here. It has been the modest function of the present review merely to carry the reader to the point where simple mathematical methods begin to be inadequate; this point having now been reached, the discussion may appropriately be brought to a close.

## Bibliography.

ABEGG, R.: Untersuchungen über Diffusion in wässerigen Salzlösungen. Z. physik. Chem. **11**, 248—264 (1893).

— u. E. BOSE: Über den Einfluß gleichioniger Zusätze auf die elektromotorische Kraft von Konzentrationsketten und auf die Diffusionsgeschwindigkeit; Neutralsalzwirkungen. Ebenda **30**, 545—555 (1899).

ADAIR, G. S.: The penetration of electrolytes into gels. II. The application of FOURIER's linear diffusion law. Biochemic. J. **14**, 762—779 (1920).

ADOLPH, E. F.: The initial rates of swelling of isolated muscle and their relation to the osmotic exchanges within the frog. Amer. J. Physiol. **96**, 598—612 (1931).

AFFONSKY, S. I.: Über die Frage des Einflusses der Lipoide auf die Diffusion der Säuren und Alkalien in Gallerten. Biochem. Z. **195**, 387—395 (1928).

ANDREWS, D. H. and J. JOHNSTON: The rate of absorption of water by rubber. J. amer. chem. Soc. **46**, 640—650 (1924).

ARRHENIUS, S.: Untersuchungen über Diffusion von in Wasser gelösten Stoffen. Z. physik. Chem. **10**, 51—95 (1892).

AUERBACH, R.: Beiträge zur Meßmethodik und zur Theorie der Diffusionsmessung gefärbter Stoffe. Kolloid-Z. **35**, 202—215 (1924).

BÄRLUND, H.: Permeabilitätsstudien an Epidermiszellen von *Rhoeo discolor*. Acta bot. fenn. **5**, 9—117 (1929).

BARCROFT, J., C. A. BINGER, A. V. BOCK, J. H. DOGGART, H. S. FORBES, G. HARROP, J. C. MEAKINS and A. C. REDFIELD: Observations upon the effect of high altitude on the physiological processes of the human body, carried out in the Peruvian Andes, chiefly at Cerro de Pasco. Philos. trans. roy. Soc. Lond. B **211**, 351—480 (1922).

— A. COOKE, H. HARTRIDGE, T. R. PARSONS and W. PARSONS: The flow of oxygen through the pulmonary epithelium. J. of Physiol. **53**, 450—472 (1920).

BARNES, C.: Diffusion through a membrane. Physics **5**, 4—8 (1934).

BAUER, H.: Bemerkung zu PH. FRANKs Abhandlung „Über die Fortpflanzungsgeschwindigkeit der Diffusion". Physik. Z. **20**, 339—340 (1919).

BECHHOLD, H. u. J. ZIEGLER: Niederschlagsmembranen in Gallerte und die Konstitution der Gelatinegallerte. Ann. Physik (4) **20**, 900—918 (1906a).

— — Die Beeinflußbarkeit der Diffusion in Gallerten. Z. physik. Chem. **56**, 105—121 (1906b).

BECKING, L. G. M. B., H. BAKHUYZEN and H. HOTELLING: The physical state of protoplasm. Verh. Akad. Wetensch. Amsterd. Afd. Natuurkunde **25**, 5—53 (1928).

BEHN, U.: Über wechselseitige Diffusion von Elektrolyten in verdünnten wässerigen Lösungen, insbesondere über Diffusion gegen das Concentrationsgefälle. Wiedemanns Ann., N. F. **62**, 54—67 (1897).

BERTHOLLET, C. L.: Essai de statique chimique. Paris 1803.

BOHR, C.: Definition und Methode zur Bestimmung der Invasions- und Evasionskoeffizienten bei der Auflösung von Gasen in Flüssigkeiten. Werthe der genannten Constanten sowie der Absorptionscoefficienten der Kohlensäure bei Auflösung in Wasser und in Chlornatriumlösungen. Wiedemanns Ann. **68**, 500—525 (1899).

BOLTZMANN, L.: Zur Integration der Diffusionsgleichung bei variabeln Diffusionscoefficienten. Ebenda **53**, 959—964 (1894).

BOSE, E.: Beitrag zur Diffusionstheorie. Z. physik. Chem. **29**, 658—660 (1899).

BRILLOUIN, L.: Diffusion de granules animés d'un mouvement brownien. Ann. Chim. et Phys. (8) **27**, 412—423 (1912).

BRINTZINGER, H.: Beiträge zur Kenntnis der Dialyse. I. Mitteilung. Das Abklingungsgesetz der Dialyse. Z. anorg. u. allg. Chem. **168**, 145 bis 149 (1927a).

— Beiträge zur Kenntnis der Dialyse. II. Mitteilung. Der Verlauf und die Geschwindigkeit der Dialyse eine Funktion der spezifischen Oberfläche. Ebenda **168**, 150—153 (1927b).

— Die Verwendung des Dialysenkoeffizienten zur Bestimmung des Molekulargewichts. Naturwiss. **18**, 354—355 (1930).

— u. W. BRINTZINGER: Die Bestimmung des Molekulargewichts aus dem Dialysenkoeffizienten. Z. anorg. u. allg. Chem. **196**, 33—43 (1931a).

— — Zur Kenntnis des Systems Elektrolyt-Wasser. Die Verteilung der Ionen eines Salzpaares bei der Diffusion durch eine Membran. Ebenda **196**, 61—64 (1931b).

— K. MAURER u. J. WALLACH: Messungen mit Hilfe der Dialysen-Methode. Untersuchungen über das Molekulargewicht und den thermischen Abbau des Inulins und Inulans in wäßriger Lösung. Ber. dtsch. chem. Ges. B. **65**, 988—995 (1932).

— u. B. TROEMER: Beiträge zur Kenntnis der Dialyse. III. Mitteilung. Der Temperaturkoeffizient der Dialyse. Z. anorg. u. allg. Chem. **172**, 426—428 (1928).

— — Messungen mit Hilfe der Dialysenmethode. Beiträge zur Kenntnis des Systems Elektrolyt-Wasser. Ebenda **184**, 97—120 (1929).

BROOKS, S. C.: The effect of light on the permeability of lecithin. Science (N. Y.) **61**, 214—215 (1925).

— and M. M. BROOKS: The rate of penetration of dyes into Valonia, with special reference to solubility theories of permeability. J. cell. a. comp. Physiol. **2**, 53—73 (1932).

BROWN, H. T.: Some recent work on diffusion. Nature (Lond.) **64**, 171—174, 193—195 (1901).

— The principles of diffusion, their analogies and applications. J. chem. Soc. Lond. **113**, 559—585 (1918).

— and F. ESCOMBE: Static diffusion of gases and liquids in relation to the assimilation of carbon and translocation in plants. Philos. trans. roy. Soc. Lond. B **193**, 223—291 (1900).

BROWN, R.: A brief account of microscopical observations made in the months of June, July and August 1827 on the particles contained in the pollen of plants; and on the general existence of active molecules in organic and inorganic bodies. Philosophic. Mag. **4**, 161—173 (1828).

BRÜCKE, E.: Beiträge zur Lehre von der Diffusion tropfbarflüssiger Körper durch poröse Scheidewände. Poggendorffs Ann. **58**, 77—94 (1843).

BRUINS, H. R.: Die Diffusion kolloider Teilchen. I. Abnorm große Diffusionsgeschwindigkeiten in hydrophilen Solen. Kolloid-Z. **54**, 265—272 (1931a).

— Die Diffusion kolloider Teilchen. II. Ein neuer Ioneneffekt in hydrophilen Solen. Ebenda **54**, 272—278 (1931b).

BRUINS, H. R.: Die Diffusion kolloider Teilchen. III. Nähere Untersuchung, Deutung und Bedeutung der abnorm großen Diffusionsgeschwindigkeiten und des neuen Ioneneffektes in hydrophilen Solen. Kolloid-Z. **57**, 152—166 (1931c).

— Die Diffusion kolloider Teilchen. IV. Der Einfluß der Ladung auf die Diffusionsgeschwindigkeit und auf deren Änderung während der Koagulation. Ebenda **59**, 263—266 (1932).

BURRAGE, L. J.: The diffusion of sodium chloride in aqueous solutions. J. physic. Chem. **36**, 2166—2174 (1932).

BURTON, E. F. and E. BISHOP: The law of distribution of particles in colloidal solution. Proc. roy. Soc. Lond. **100**, 414—419 (1922).

— and J. E. CURRIE: The distribution of colloidal particles. Philosophic. Mag. (6) **47**, 721—724 (1924).

BYERLY, W. E.: An Elementary Treatise on FOURIER'S Series and Spherical, Cylindrical, and Ellipsoidal Harmonics. Boston 1893.

CALUGAREANU, M. and V. HENRI: Diffusion des matières colorantes dans la gélatine et dans l'eau. C. r. Soc. Biol. Paris **53**, 579—580 (1901).

CARLSON, T.: The diffusion of oxygen in water. J. amer. chem. Soc. **33**, 1027—1032 (1911a).

— On the diffusion of oxygen and carbon dioxide in water. Meddel. fran Vet.-Akad. Nobelinstitut **2**, Nr 6, 1—8 (1911b).

CARSLAW, H. S.: Introduction to the Mathematical Theory of the Conduction of Heat in Solids. London 1921.

CHABRY, P.: Procédé nouveau pour étudier la diffusion des acides. J. Physique (2) **7**, 114—122 (1888).

CHAUDESAIGUES, M.: Le mouvement brownien et la formule d'EINSTEIN. C. r. Acad. Sci. Paris **147**, 1044—1046 (1908).

CLACK, B. W.: On the coefficient of diffusion. Philosophic. Mag. (6) **16**, 863—879 (1908a).

— On the coefficient of diffusion. Proc. physic. Soc. Lond. **21**, 374—392 (1908b).

— On the temperature coefficient of diffusion. Ebenda **24**, 40—49 (1911).

— (4) On the coefficient of diffusion in dilute solutions. Ebenda **27**, 56—69 (1914).

— On diffusion in liquids. Ebenda **29**, 49—58 (1916).

— On the coefficient of diffusion of certain saturated solutions. Ebenda **33**, 259—265 (1921).

— On the study of diffusion in liquids by an optical method. Ebenda **36**, 313—335 (1924).

COHEN, E. u. H. R. BRUINS: Ein Präzisionsverfahren zur Bestimmung von Diffusionskoeffizienten in beliebigen Lösungsmitteln. Z. physik. Chem. **103**, 349—403 (1923a).

— — Über die Gültigkeit des STOKES-EINSTEINschen Gesetzes für diffundierende Moleküle. Ebenda **103**, 404—450 (1923b).

— — Ein Präzisionsverfahren zur Bestimmung von Diffusionskoeffizienten in beliebigen Lösungsmitteln, II. Ebenda **113**, 157—159 (1924).

COLLANDER, R. u. H. BÄRLUND: Permeabilitätsstudien an *Chara ceratophylla*. II. Die Permeabilität für Nichtelektrolyte. Acta bot. fenn. **11**, 1—114 (1933).

DABROWSKI, S.: Méthode de fractionnement par diffusion; son application à l'étude des solutions colloïdales (1ère partie). Bull. internat. Acad. Sci. Cracovie A' **1912**, 485—526.

DAVIES, R. J.: The diffusion of electrolytes. Philosophic. Mag. (7) **15**, 489 bis 511 (1933).

DAVIS, H. S. and G. S. CRANDALL: The rôle of the liquid stationary film in batch absorptions of gases. I. Absorption involving no irreversible chemical reactions. J. amer. chem. Soc. **52**, 3757—3768 (1930).

DAYNES, H. A.: The process of diffusion through a rubber membrane. Proc. roy. Soc. Lond. A **97**, 286—307 (1920).

DES COUDRES, TH.: Formel für Diffusionsvorgänge in einem Cylinder von endlicher Länge bei Einwirkung der Schwere. Wiedemanns Ann., N. F. **55**, 213—219 (1895).

DIRKEN, M. N. J. u. H. W. MOOK: Geschwindigkeit der Kohlensäureaufnahme durch Flüssigkeiten. Biochem. Z. **219**, 452—462 (1930).

DOMBROWSKY, S.: Régime de concentrations établi par la diffusion latérale dans un courant de convection. C. r. Acad. Sci. Paris **180**, 1581—1583 (1925).

DONNAN, F. G.: Concerning the applicability of thermodynamics to the phenomena of life. J. gen. Physiol. **8**, 685—688 (1927).

DUMANSKI, A.: Über die Diffusion im kolloiden Medium. Kolloid-Z. **3**, 210—212 (1908).

DUMMER, E.: Diffusion organischer Substanzen in organischen Lösungsmitteln und Prüfung der STOKESschen Formel. Z. anorg. u. allg. Chem. **109**, 31—51 (1919).

DUTROCHET, R.: Nouvelles observations sur l'endosmose et l'exosmose et sur la cause de ce double phénomène. Ann. Chim. et Phys. **35**, 393—400 (1827).

EAGLE, A.: A Practical Treatise on FOURIER's Theorem and Harmonic Analysis. London 1925.

EDGAR, G. and S. H. DIGGS: The diffusion of iodine in potassium iodide solutions. J. amer. chem. Soc. **38**, 253—259 (1916).

EGGLETON, G. P., P. EGGLETON and A. V. HILL: The coefficient of diffusion of lactic acid through muscle. Proc. roy. Soc. Lond. B **103**, 620—628 (1928).

EGGLETON, P.: The diffusion of creatine and urea through muscle. J. of Physiol. **70**, 294—300 (1930).

EINSTEIN, A.: Über die von der molekularkinetischen Theorie der Wärme geforderte Bewegung von in ruhenden Flüssigkeiten suspendierender Teilchen. Ann. Physik. (4) **17**, 549—560 (1905).

— Elementare Theorie der BROWNschen Bewegung. Z. f. Elektrochem. **14**, 235—239 (1908).

— Untersuchungen über die Theorie der BROWNschen Bewegung. OSTWALD's Klassiker der exakten Wissenschaften Nr. 199. Leipzig 1922.

EPPINGER, H. u. W. BRANDT: Über die Beeinflußbarkeit der Diffusion I von Salzen durch Gallerten, II von Gasen durch Membranen. Biochem. Z. **249**, 11—20 (1932).

EULER, H.: Über die Beweglichkeit von nicht dissociirten und dissociirten Molekülen. Wiedemanns Ann., N. F. **63**, 273—277 (1897).

EXNER, F.: Über den Durchgang der Gase durch Flüssigkeitslamellen. Poggendorffs Ann. **155**, 321—336, 443—464 (1875).

FENN, W. O.: The oxygen consumption of frog nerve during stimulation. J. gen. Physiol. **10**, 767—779 (1927).

FICK, A.: Über Diffusion. Poggendorffs Ann. **94**, 59—86 (1855).

FISCHER, F. P.: Über die Diffusion des Hämoglobins in kolloide und molekulardisperse Lösungen. Kolloid-Z. **57**, 166—173 (1931).

FISCHER, N. W.: Über die Wiederherstellung eines Metalls durch ein anderes, und über die Eigenschaft der thierischen Blase, Flüssigkeiten durch sich hindurch zu lassen, und sie in einigen Fällen anzuheben. Gilberts Ann. **12**, 289—307 (1822).

FRANK, P.: Über die Fortpflanzungsgeschwindigkeit der Diffusion. Physik. Z. **19**, 516—520 (1918).
FRANKE, G.: Zur Diffusion von Alkoholen. Ann. Physik (5) **14**, 675—682 (1932).
FREUNDLICH, E.: Zur Prüfung der allgemeinen Relativitätstheorie. Naturwiss. **7**, 629—636 (1919).
FRICKE, R.: Über eine allgemeine Methode zur exakten Untersuchung von Diffusionsvorgängen in Gallerten. Z. Elektrochem. **31**, 430—435 (1925).
FRIEDMAN, L.: Diffusion of non-electrolytes in gelatin gels. J. amer. chem. Soc. **52**, 1305—1310 (1930a).
— Structure of agar gels from studies of diffusion. Ebenda **52**, 1311 bis 1314 (1930b).
— and E. O. KRAEMER: The structure of gelatin gels from studies of diffusion. Ebenda **52**, 1295—1304 (1930).
FÜRTH, O., H. BAUER u. H. PIESCH: Untersuchungen über Diffusionsvorgänge in Gallerten. III. Über die Beziehungen des Diffusionsweges zum Diffusionskoeffizienten und seine Abhängigkeit von der Konzentration des Indicators. Biochem. Z. **100**, 29—63 (1919).
— u. F. BUBANOVIĆ: Untersuchungen über die Diffusion von Elektrolyten in Gallerten. I. Über die Abhängigkeit des Diffusionsweges von der Konzentration. Ebenda **90**, 265—287 (1918a).
— — Untersuchungen über Diffusionsvorgänge in Gallerten. II. Über die Abhängigkeit des Diffusionsvermögens von der Ionenbeweglichkeit sowie von der Hydratation und Polymerisation der Moleküle. Ebenda **92**, 139—169 (1918b).
FÜRTH, R.: Einige Untersuchungen über BROWNsche Bewegung an einem Einzelteilchen. Ann. Physik (4) **53**, 177—213 (1917).
— Diffusionsversuche an Lösungen. Physik. Z. **26**, 719—722 (1925).
— Über ein Problem der Diffusion im Schwerfelde. Z. Physik **40**, 351—363 (1927a).
— Über Diffusion im Schwerfelde. Ebenda **45**, 83—85 (1927b).
— Wärmeleitung und Diffusion. RIEMANN-WEBER: Die Differential- und Integralgleichungen der Mechanik und Physik, II. Abschnitt, Teil 2, S. 177—266. Braunschweig 1927c.
— Zur physikalischen Chemie der Farbstoffe. IV. Eine neue Methode zur exakten Bestimmung des Dispersitätsgrades der Farbstofflösungen. Kolloid Z. **41**, 300—304 (1927d).
— Diffusion ohne Scheidewände. AUERBACH und HORTS Handbuch der physiologischen und technischen Mechanik, Bd. 7, S. 635—704. 1931a.
— Disperse Systeme und BROWNsche Bewegung. Ebenda Bd. 7, S. 198—238. 1931b.
— Untersuchungen über Diffusion in Flüssigkeiten. I. Vorbemerkung. Z. Physik **79**, 275—279 (1932).
— u. E. ULLMANN: Zur physikalischen Chemie des Farbstoffes. V. Untersuchungen über den Dispersitätsgrad von Farbstofflösungen. Kolloid-Z. **41**, 304—310 (1927).
FUJITA, A.: Untersuchungen über elektrische Erscheinungen und Ionendurchlässigkeit von Membranen. VIII. Mitteilung: Die Permeabilität der getrockneten Kollodiummembran für Nichtelektrolyte. Biochem. Z. **170**, 18—29 (1926).
GELLHORN, E.: Das Permeabilitätsproblem. Berlin 1929.
GERARD, R. W.: Studies on nerve metabolism. II. Respiration in oxygen and nitrogen. Amer. J. Physiol. **82**, 381—404 (1927).
— Oxygen diffusion into cells. Biol. Bull. Mar. biol. Labor. Wood's Hole **60**, 245—268 (1931).

Gerlach, B.: Experimentelle Untersuchungen über die Diffusion von Flüssigkeiten. Ann. Physik (5) **10**, 437—459 (1931).

Goldschmidt, S. and A. Hunsberger: The diffusion rate of ions as affected by the presence of other ions in a solution. Amer. J. Physiol. **90**, 362 (1929).

Graham, E. A. and H. T. Graham: Retardation by sugars of diffusion of acids in gels. J. amer. chem. Soc. **40**, 1900—1917 (1918).

Graham, J. C.: Über die Diffusion von Salzen in Lösung. Z. physik. Chem. **50**, 257—272 (1904).

— Diffusion von Salzen in Lösung, II. Ebenda **59**, 691—696 (1907).

Graham, T.: On the diffusion of liquids. Philos. trans. roy. Soc. Lond. **140**, 1—46 (1850a).

— Supplementary observations on the diffusion of liquids. Ebenda **140**, 805—836 (1850b).

— Additional observations on the diffusion of liquids (Third memoir). Ebenda **141**, 483—494 (1851a).

— Über die Diffusion von Flüssigkeiten. Liebigs Ann. **77**, 56—89, 129—160 (1851b).

— Liquid diffusion applied to analysis. Philos. trans. roy. Soc. Lond. **151**, 183—224 (1861).

— Liquid diffusion applied to analysis. Philosophic. Mag. (4) **23**, 204 bis 223, 290—306, 368—380 (1862a).

— Mémoire sur la diffusion moléculaire appliquée a l'analyse. Ann. Chim. et Phys. (3) **65**, 129—207 (1862b).

Griffiths, A.: Diffusive convection. Philosophic. Mag. (5) **46**, 453—465 (1898).

— A study of an apparatus for the determination of the rate of diffusion of solids dissolved in liquids. Ebenda (5) **47**, 530—539 (1899).

— On the determination, by the method of diffusive convection, of the coefficient of diffusion of a salt dissolved in water. Proc. physic. Soc. Lond. **28**, 21—34 (1916a).

— A recalculation of some work on diffusion by Dr. A. Griffiths and others. Ebenda **28**, 255—257 (1916b).

— Note on the calculation of the coefficient of diffusion of a salt at a definite concentration. Ebenda **29**, 159—162 (1917).

— J. M. Dickson and C. H. Griffiths: Determination of the coefficient of diffusion of potassium chloride by an analytical method. Ebenda **28**, 73—80 (1916).

Gróh, J. u. I. Kelp: Diffusion des Jods in reinen Lösungsmitteln und in Lösungsmittelgemischen. Z. anorg. u. allg. Chem. **147**, 321—330 (1925).

Guyer, A. u. B. Tobler: Zur Kenntnis der Geschwindigkeit der Gasexsorption von Flüssigkeiten. Helvet. chim. Acta **17**, 257—271, 550—555 (1934).

Haas-Lorentz, G. L. de: Die Brownsche Bewegung und einige verwandte Erscheinungen. Braunschweig 1913.

Hagenbach, A.: Über Diffusion von Gasen durch wasserhaltige Gelatine. Wiedemanns Ann., N. F. **65**, 673—706 (1898).

Hartley, G. S.: Theory of the velocity of diffusion of strong electrolytes in dilute solution. Philosophic. Mag. (7) **12**, 473—488 (1931).

— and C. Robinson: The diffusion of colloidal electrolytes and other charged colloids. Proc. roy. Soc. Lond. A **134**, 20—35 (1931).

Hartridge, H.: Shape of red blood corpuscles. J. of Physiol. **53**, 81 (1920).

— and F. J. W. Roughton: The rate of distribution of dissolved gases between the red blood corpuscle and its fluid environment. Part I. Preliminary experiments on the rate of uptake of oxygen and carbon monoxide by sheep's corpuscles. Ebenda **62**, 232—242 (1927).

HARVEY, E. N.: The oxygen consumption of luminous bacteria. J. gen. Physiol. **11**, 469—475 (1928).
HASKELL, R.: The effect of concentration and ionization on the rates of diffusion of salts in aqueous solution. Physic. Rev. **27**, 145—182 (1908).
HATSCHEK, E.: Die Diffusions-Geschwindigkeit in Gelatinegelen als Funktion der Viskosität des Dispersionsmittels. Kolloid-Z. **60**, 273—276 (1932).
HAUSMANINGER, V.: Über die Veränderlichkeit des Diffusionscoefficienten zwischen Kohlensäure und Luft. Sitzgsber. Akad. Wiss. Wien, Math.-naturwiss. Kl. II **86**, 1073—1089 (1882).
HEIMBRODT, FR.: Diffusionskoeffizienten in Abhängigkeit von der Konzentration, bestimmt mit Hilfe gekrümmter Lichtstrahlen. Ann. Physik (4) **13**, 1028—1043 (1904).
HERTZ, G.: Ein neues Verfahren zur Trennung von Gasgemischen durch Diffusion. Physik. Z. **23**, 433—434 (1922).
— Über Trennung von Gasgemischen durch Diffusion in einem strömenden Gase. Z. Physik **19**, 35—42 (1923).
HERZOG, R. O.: Über die Diffusion der Kolloide, I. Kolloid-Z. **2**, 1—2 (1907a).
— Diffusion von Kolloiden. Z. Elektrochem. **13**, 533—541 (1907b).
— u. H. KASARNOWSKI: Über die Diffusion von Kolloiden, II. Biochem. Z. **11**, 172—176 (1908).
— A. POLOTZKY: Die Diffusion einiger Farbstoffe. Z. physik. Chem. **87**, 449—489 (1914).
HEVESY, G. v.: Die Valenz der Radioelemente. Physik. Z. **14**, 49—62 (1913a).
— The valency of the radio elements. Philosophic. Mag. (6) **25**, 390—414 (1913b).
HILL, A. V.: Note on the use of the experimental method described in the preceding paper. Proc. Cambridge philos. Soc. **15**, 387—389 (1910).
— The diffusion of oxygen and lactic acid through tissues. Proc. roy. Soc. Lond. B **104**, 39—96 (1928).
HÖBER, R.: Physikalische Chemie der Zelle und der Gewebe, 6. Aufl. Leipzig 1926.
HÖFLER, K.: Über Eintritts- und Rückgangsgeschwindigkeit der Plasmolyse und eine Methode zur Bestimmung der Wasserpermeabilität des Protoplasten. Jb. Bot. **73**, 300—350 (1930).
HOELTZENBEIN, FR.: Die Methode von H. F. WEBER zur Bestimmung des Diffusionskoeffizienten von Salzlösungen. Z. Physik **24**, 1—9 (1924).
HOFFMANN, G.: Diffusion von Thorium X. Ann. Physik (4) **21**, 239—269 (1906).
HOPPE-SEYLER, F.: Medicinisch-chemische Untersuchungen. Berlin 1866.
HUBER, B. u. K. HÖFLER: Die Wasserpermeabilität des Protoplasmas. Jb. Bot. **73**, 351—511 (1930).
HÜFNER, G.: Über die Bestimmung des Diffusionscoefficienten einiger Gase für Wasser. Wiedemanns Ann., N. F. **60**, 134—168 (1897).
— Über die Diffusion von Gasen durch Wasser und durch Agargallerte. Z. physik. Chem. **27**, 227—249 (1898).
INGERSOLL, L. R. and O. J. ZOBEL: An Introduction to the Mathematical Theory of Heat Conduction. Boston 1913.
INGRAHAM, R. C., C. LOMBARD and M. B. VISSCHER: The characteristics of ultrafiltrates of plasma. J. gen. Physiol. **16**, 637—655 (1933).
JABLCZYNSKI, K.: Diffusion a travers les membranes. J. Chim. physique **7**, 117—128 (1909).
JACOBS, M. H.: Permeability of the cell to diffusing substances. Cowdry: General Cytology, p. 99—164. Chicago 1924.

JACOBS, M. H.: The exchange of material between the erythrocyte and its surroundings. Harvey Lectures 22, 146—164 (1927).
— Osmotic properties of the erythrocyte. III. The applicability of osmotic laws to the rate of hemolysis in hypotonic solutions of non-electrolytes. Biol. Bull. Mar. biol. Labor. Wood's Hole **62**, 178—194 (1932).
— The relation between cell volume and penetration of a solute from an isosmotic solution. J. cell. a. comp. Physiol. **3**, 29—43 (1933a).
— Volume changes of cells in solutions containing both penetrating and non-penetrating solutes, and their relation to the 'permeability ratio'. Ebenda **3**, 121—129 (1933b).
— The simultaneous measurement of cell permeability to water and to dissolved substances. Ebenda **4**, 427—444 (1933c).
— The quantitative measurement of the permeability of the erythrocyte to water and to solutes by the hemolysis method. Ebenda **4**, 161—183 (1934).
— and D. R. STEWART: A simple method for the quantitative measurement of cell permeability. Ebenda **1**, 71—82 (1932).
JANDER, G. u. H. SCHULZ: Versuche über die Verwendung des Diffusionskoeffizienten zur Bestimmung der Molekulargröße schwerer, amphoterer Oxydhydrate in alkalischer Lösung. Kolloid-Z. **36** (ZSIGMONDY Festschrift), 109—118 (1925a).
— Über amphotere Oxydhydrate, deren alkalische Lösungen und feste Salze (Isopolysäuren und isopolysaure Salze). II. Mitteilung. Z. anorg. u. allg. Chem. **144**, 225—247 (1925b).
JERICHAU, E. B.: Über das Zusammenströmen flüssiger Körper, welche durch poröse Lamellen getrennt sind. Poggendorffs Ann. **34**, 613—627 (1835).
JOHANNISJANZ, A.: Über die Diffusion der Flüssigkeiten. Wiedemanns Ann., N. F. **2**, 24—47 (1877).
JOLLY, PH.: Experimental-Untersuchungen über Endomsose. Poggendorffs Ann. **78**, 261—271 (1849).
KAWALKI, W.: Untersuchungen über die Diffusionsfähigkeit einiger Elektrolyte in Alkohol. Ein Beitrag zur Lehre von der Constitution der Lösungen. Wiedemanns Ann., N. F. **52**, 166—190, 300—327 (1894).
— Die Abhängigkeit der Diffusionsfähigkeit von der Anfangsconcentration bei verdünnten Lösungen. Ebenda **59**, 637—651 (1896).
KLEBS, G.: Beiträge zur Physiologie der Pflanzenzelle. Untersuch. bot. Inst. Tübingen **2**, 489—568 (1888).
KLEMM, K. and L. FRIEDMAN: The structure of cellulose acetate gels from studies of diffusion. J. amer. chem. Soc. **54**, 2637—2642 (1932).
KRIJGSMAN, B. J.: Neuere Ansichten über die Permeabilität von nichtlebenden und lebenden Membranen. Erg. Biol. **9**, 292—357 (1932).
KROGH, A.: Some experiments on the invasion of oxygen and carbonic oxide into water. Skand. Arch. Physiol. (Berl. u. Lpz.) **23**, 224—235 (1910).
— The rate of diffusion of gases through animal tissues, with some remarks on the coefficient of invasion. J. of Physiol. **52**, 391—408 (1919a).
— The number and distribution of capillaries in muscles with calculations of the oxygen pressure head necessary for supplying the tissue. Ebenda **52**, 409—415 (1919b).
— and M. KROGH: On the rate of diffusion of carbonic oxide into the lungs of man. Skand. Arch. Physiol. (Berl. u. Lpz.) **23**, 236—247 (1910).
KROGH, M.: The diffusion of gases through the lungs of man. J. of Physiol. **49**, 271—300 (1915).

KRUGER, D. u. H. GRUNSKY: Über die Diffusion von Stoffen, die Abweichungen vom FICKschen Gesetz zeigen. Z. physik. Chem. A **150**, 115 bis 134 (1930).

KUNITZ, M., M. L. ANSON and J. H. NORTHROP: Molecular weight, molecular volume, and hydration of proteins in solution. J. gen. Physiol. **17**, 365—373 (1934).

LAMM, O.: Zur Bestimmung von Konzentrationsgradienten mittels gekrümmter Lichtstrahlen. Eine neue Beobachtungsmethode. Z. physik. Chem. A **138**, 313—331 (1928).

LANDIS, E. M.: Micro-injection studies of capillary permeability. II. The relation between capillary pressure and the rate at which fluid passes through the walls of single capillaries. Amer. J. Physiol. **82**, 217—238 (1927).

LANGEVIN, P.: Sur la theorie du mouvement brownien. C. r. Acad. Sci. Paris **146**, 530—533 (1908).

LEITCH, J. L.: The water exchanges of living cells. I. The normal permeability of the eggs of some marine invertebrates. Univ. California Pub. Zool. **36**, 127—140 (1931).

LEPESCHKIN, W. W.: Über den Turgordruck der vakuolisierten Zellen. Ber. dtsch. bot. Ges. **26a**, 198—214 (1908).

LEWIS, G. N. and M. RANDALL: Thermodynamics and the Free Energy of Chemical Substances. New York 1923.

LEWIS, W. K. and W. G. WHITMAN: Principles of gas absorption. Ind. Engng. Chem. **16**, 1215—1220 (1924).

LILLIE, R. S.: Increase of permeability to water following normal and artificial activation in sea-urchin eggs. Amer. J. Physiol. **40**, 249—266 (1916).

LITTLEWOOD, T. H.: On the diffusion of solutions. Proc. physic. Soc. Lond. **34**, 71—76 (1922).

LONGSWORTH, L. G.: The theory of diffusion in cell models. J. gen. Physiol. **17**, 211—235 (1933).

LOSCHMIDT, J.: Experimental-Untersuchungen über die Diffusion von Gasen ohne poröse Scheidewände. Sitzgsber. Akad. Wiss. Wien, Math.- naturwiss. Kl. II **61**, 367—380 (1870a).

— Experimental-Untersuchungen über die Diffusion von Gasen ohne poröse Scheidewände. Ebenda II **62**, 468—478 (1870b).

LUCKÉ, B. and M. McCUTCHEON: The living cell as an osmotic system and its permeability to water. Physiol. Rev. **12**, 68—139 (1932).

— H. K. HARTLINE and M. McCUTCHEON: Further studies on the kinetics of osmosis in living cells. J. gen. Physiol. **14**, 405—419 (1931).

LUDWIG, C.: Über die endosmotischen Äquivalente und die endosmotische Theorie. Poggendorffs Ann. **78**, 307—326 (1849).

LUNDSGAARD, C. and S. A. HOLBØLL: Investigations into the standardization and calibration of collodion membranes, I. J. of biol. Chem. **68**, 439 bis 456 (1926).

McBAIN, J. W.: Der Mechanismus der Adsorption („Sorption") von Wasserstoff durch Kohlenstoff. Z. physik. Chem. **68**, 471—497 (1909).

— and C. R. DAWSON: Accelerated and retarded diffusion in aqueous solution. J. amer. chem. Soc. **56**, 52—56 (1934).

— — and H. A. BARKER: The diffusion of colloids and colloidal electrolytes; egg albumin; comparison with ultracentrifuge. Ebenda **56**, 1021—1027 (1934).

— and T. H. LIU: Diffusion of electrolytes, non-electrolytes and colloidal electrolytes. Ebenda **53**, 59—74 (1931).

McBain, J. W., P. J. van Rysselberge and W. A. Squance: The degree of dissociation and the ions of cadmium iodide in aqueous solution. J. physic. Chem. **35**, 999—1010 (1931).
McBain, M. E. L.: The diffusion of colloidal electrolytes; sodium oleate. J. amer. chem. Soc. **55**, 545—551 (1933).
McIver, M. A., A. C. Redfield and E. B. Benedict: Gaseous exchange between the blood and the lumen of the stomach and intestines. Amer. J. Physiol. **76**, 92—111 (1926).
McKay, A. T.: Diffusion into an infinite plane sheet subject to a surface condition, with a method of application to experimental data. Proc. physic. Soc. Lond. **42**, 547—555 (1930).
— Diffusion for the infinite plane sheet. Ebenda **44**, 17—24 (1932a).
— Absorption and classical diffusion. Trans. Faraday Soc. **28**, 721—731 (1932b).
Mack, E., jr.: Average cross-sectional areas of molecules by gaseous diffusion methods. J. amer. chem. Soc. **47**, 2468—2482 (1925).
Magistris, H.: Zur Biochemie und Physiologie organischer Phosphorverbindungen in Pflanze und Tier. IV. Mitt. Ein Beitrag zur Frage des Einflusses von tierischem und pflanzlichem Lecithin auf die Diffusion von Säuren und Alkalien in Gallerten. Biochem. Z. **253**, 81—96 (1932).
Magnus, G.: Über einige Erscheinungen der Capillarität. Poggendorffs Ann. **10**, 153—168 (1827).
Manegold, E. u. K. Solf: Über Kapillarsysteme, XIV. 1. Die Dynamik osmotischer Zellen. Kolloid-Z. **59**, 179—195 (1932).
— u. C. Stüber: Über Kapillar-Systeme XIV (2). Zur Dynamik der Plasmolyse. Erster Teil. Die mathematische Behandlung semipermeabler Protoplasten. Ebenda **63**, 316—323 (1933).
Mann, C. E. T.: The determination of coefficients of diffusion in gels by means of chemical analysis, and a comparison of results obtained with those yielded by the indicator method. Proc. roy. Soc. Lond. A **105**, 270—281 (1924).
March, H. W. and W. Weaver: The diffusion problem for a solid in contact with a stirred liquid. Physic. Rev. **31**, 1072—1082 (1928).
Marignac, M. C.: Recherches sur la diffusion simultanée de quelques sels. Ann. Chim. et Phys. (5) **2**, 546—581 (1874).
Mason, M. and W. Weaver: The settling of small particles in a fluid. Physic. Rev. **23**, 412—426 (1924).
Meyer, K.: Über die Diffusion in Gallerten. Beitr. chem. Physiol. u. Path. **7**, 392—410 (1906).
Meyerhof, O. u. W. Schulz: Über das Verhältnis von Milchsäurebildung und Sauerstoffverbrauch bei der Muskelkontraktion. Pflügers Arch. **217**, 547—573 (1927).
Miller, C. C.: The Stokes-Einstein law for diffusion in solution. Proc. roy. Soc. Lond. A **106**, 724—749 (1924).
Minami, S.: Versuche an überlebendem Carcinomgewebe. (Atmung und Glykolyse.) Biochem. Z. **142**, 334—350 (1923).
Mines, G. R.: On the relative velocities of diffusion in aqueous solution of rubidium and caesium chlorides. Proc. Cambridge philos. Soc. **15**, 381 bis 386 (1910).
Muchin, G. E. u. G. P. Faermann: Diffusionsgeschwindigkeit und Lösungsmittel. Z. physik. Chem. **121**, 180—188 (1926).
Münter, E.: Experimentelle Untersuchungen über Diffusion von Flüssigkeiten. Ann. Physik (5) **11**, 558—578 (1931).
Nell, P.: Studien über Diffusionsvorgänge wässeriger Lösungen in Gelatine. Ebenda (4) **18**, 323—347 (1905).

NERNST, W.: Zur Kinetik der in Lösung befindlichen Körper. Z. physik. Chem. **2**, 613—637 (1888).
— Theoretische Chemie vom Standpunkte der AVOGADROschen Regel und die Thermodynamik, 11. bis 15. Aufl. Stuttgart 1926.
NISTLER, A.: Dispersoidanalyse mittels eines neuen Diffusionsapparates. Kolloidchem. Beih. **28**, 296—313 (1929).
— Dispersitätsuntersuchungen an Farbstoffen. Ebenda **31**, 1—58 (1930).
— Über die Bedeutung der Dispersoidanalyse und eine Neukonstruktion des Diffusionsmikroskopes. Protoplasma (Berl.) **13**, 517—545 (1931).
NOLLET: Histoire de l'Acad. Roy. des Sciences, 1748, p. 101. Citation from A. FINDALY: Osmotic Pressure London 1919.
NORTHROP, J. H.: The kinetics of osmosis. J. gen. Physiol. **10**, 883—892 (1927a).
— Kinetics of the swelling of cells and tissues. Ebenda **11**, 43—56 (1927b).
— The permeability of dry collodion membranes, II. Ebenda **12**, 435—461 (1929).
— Crystalline pepsin. I. Isolation and tests of purity. Ebenda **13**, 739 bis 766 (1930).
— and M. L. ANSON: A method for the determination of diffusion constants and the calculation of the radius and weight of the hemoglobin molecule. Ebenda **12**, 543—554 (1929).
— and M. KUNITZ: Crystalline trypsin. II. General properties. Ebenda **16**, 295—311 (1932).
OBERMAYER, A. v.: Über die Abhängigkeit des Diffusionscoefficienten der Gase von der Temperatur. Sitzgsber. Akad. Wiss. Wien, Math.-naturwiss. Kl. II **81**, 1102—1127 (1880).
— Versuche über Diffusion von Gasen. Ebenda II **85**, 147—168, 748—761 (1882); **87**, 188—263 (1883); **96**, 546—577 (1887).
ÖHOLM, L. W.: Über die Hydrodiffusion der Elektrolyte. Z. physik. Chem. **50**, 309—349 (1905).
— Die freie Diffusion der Nichtelektrolyte. I. Über die Hydrodiffusion einiger optisch aktiver Substanzen. Ebenda **70**, 378—407 (1910).
— A new method for determining the diffusion of dissolved substance. Meddl. från. Vet.-Akad. Nobelinstitut **2**, Nr 22, 1—23 (1912a).
— Die freie Diffusion der Nichtelektrolyte. Über die Hydrodiffusion einiger organischer Substanzen. Ebenda **2**, Nr 23, 1—52 (1912b).
— Investigation of the diffusion of some organic substances in ethyl alcohol. Ebenda **2**, Nr 24, 1—34 (1912c).
— Contribution to the knowledge of the dependence of the diffusion on the viscosity of the solvent. Ebenda **2**, Nr 26, 1—21 (1913a).
— The diffusion of an electrolyte in gelatine. Ebenda **2**, Nr 30, 1—8 (1913b).
ONSAGER, L. and R. M. FUOSS: Irreversible processes in electrolytes. Diffusion, conductance, and viscous flow in arbitrary mixtures of strong electrolytes. J. physic. Chem. **36**, 2689—2778 (1932).
OSBORNE, W. A. and L. C. JACKSON: Counter diffusion in aqueous solution. Biochemic. J. **8**, 246—249 (1914).
OSTERHOUT, W. J. V.: Permeability in large plant cells and in models. Erg. Physiol. **35**, 967—1021 (1933).
OSTWALD, WO. u. A. QUAST: Über die Änderungen physikalisch-chemischer Eigenschaften im Übergangsgebiet zwischen Kolloiden und molekulardispersen Systemen, I. Kolloid-Z. **48**, 83—95 (1929).
PEKAREK, J.: Absolute Viskositätsmessung mit Hilfe der BROWNschen Molekularbewegung, I. Prinzip der Methode, Voraussetzungen, Fehlerquellen der Messungen. Protoplasma (Berl.) **10**, 510—532 (1930a).

PEKAREK, J.: Absolute Viskositätsmessung mit Hilfe der BROWNschen Molekularbewegung. II. Viskositätsbestimmung des Zellsaftes der Epidermiszellen von *Allium cepa* und des Amoeben-Protoplasmas. Protoplasma (Berl.) **11**, 19—48 (1930b).
— Absolute Viskositätsmessungen mit Hilfe der BROWNschen Molekularbewegung, III. a) Viskositätsmessungen an destilliertem Wasser, b) Viskositätsmessungen an Glyzerin-Wassergemischen, c) Viskositätsmessungen des Zellsaftes der Protonemazellen von *Leptobryum piriforme*. Ebenda **13**, 637—665 (1931).
— Absolute Viskositätsmessungen mit Hilfe der BROWNschen Molekularbewegung. IV. Mitt. Plasmaviskositätsmessungen an Rhizoiden von *Chara fragilis* DESV. Ebenda **17**, 1—24 (1933a).
— Absolute Viskositätsmessungen mit Hilfe der BROWNschen Molekularbewegung. V. Mitt. Plasmolyse und Zellsaftviskosität. Ebenda **18**, 1—53 (1933b).
— Absolute Viskositätsmessungen mit Hilfe der BROWNschen Molekularbewegung. VI. Mitt. Der Einfluß der Temperatur auf die Zellsaftviskosität. Ebenda **20**, 251—278 (1933c).
— Absolute Viskositätsmessungen mit Hilfe der BROWNschen Molekularbewegung. VII. Mitt. Der Einfluß des Lichtes auf die Zellsaftviskosität. Ebenda **20**, 359—375 (1933c).
PERRIN, J.: L'agitation moléculaire et le mouvement brownien. C. r. Acad. Sci. Paris **146**, 967—970 (1906a).
— Grandeur des molecules et charge de l'electron. Ebenda **147**, 594 bis 596 (1908b).
— Mouvement brownien et réalité moléculaire. Ann. Chim. et Phys. (8) **18**, 4—114 (1909).
— Les Atomes. 5. Edition, 1914.
PFEFFER, W.: Osmotische Untersuchungen. Leipzig 1877.
PICKERING, S. U.: Some experiments on the diffusion of substances in solution. Philosophic. Mag. (5) **35**, 127—134 (1893).
POISSON: Note sur des effets qui peuvent être produits par la capillarité et l'affinité des substances hétérogènes. Ann. Chim. et Phys. **35**, 98 bis 102 (1827).
PONDER, E.: The shape of the mammalian erythrocyte and its respiratory function. J. gen. Physiol. **9**, 197—204 (1925).
— The shape of the mammalian erythrocyte and its respiratory function, II. Ebenda **9**, 625—629 (1926).
PORTER, A. W. and J. J. HEDGES: The law of distribution of particles in colloidal suspensions with special reference to PERRIN's investigations. Ph losophic. Mag. (6) **44**, 641—651 (1922).
PRINGSHEIM, N.: Über chemische Niederschläge in Gallerte. Jb. Bot. **28**, 1—38 (1895a).
— Über chemische Niederschläge in Gallerte. Z. physik. Chem. **17**, 473—504 (1895b).
PROCOPIU, ST.: Sur une nouvelle méthode pour la mesure du coefficient de diffusion des électrolytes. Ann. Physique **9**, 96—112 (1918).
RAMSTEDT, E.: Sur la diffusion de l'émanation du radium dans l'eau. Meddel. från Vet.-Akad. Nobelinstitut **5**, Nr 5, 1—14 (1919).
RASHEVSKY, N.: Note on the mathematical theory of oxygen consumption at low oxygen pressures. Protoplasma (Berl.) **20**, 125—130 (1933).
REFORMATSKY, S.: Über die Geschwindigkeit chemischer Reaktionen in Gallerte. Z. physiol. Chem. **7**, 34—35 (1891).
RICKETTS, V. L. and J. L. CULBERTSON: Diffusion in alkaline copper systems. J. amer. chem. Soc. **53**, 4002—4008 (1931).

RIECKE, E.: Molekulartheorie der Diffusion und Elektrolyse. J. physic. Chem. **6**, 564—572 (1890).
RONA, E.: Diffusionsgröße und Atomdurchmesser der Radiumemanation. Ebenda **92**, 213—218 (1918).
— Diffusionsgröße und Ionenbeweglichkeit des Kobalt- und Nickelions. Ebenda **95**, 62—65 (1920).
ROUGHTON, F. J. W.: Diffusion and chemical reaction velocity as joint factors in determining the rate of uptake of oxygen and carbon monoxide by the red blood corpuscle. Proc. roy. Soc. Lond. B **111**, 1—36 (1932).
RUNNSTRÖM, J.: Untersuchungen über die Permeabilität des Seeigeleies für Farbstoffe. Ark. Zool. (schwed.) **7** (13), 1—17 (1911).
SAMEC, M., L. KNOP u. Z. PANOVIČ: Osmose und Diffusion einiger Pflanzenkolloide. (Stärkesubstanzen — Ligninsulfosäure — Humate.) Kolloid-Z. **59**, 266—278 (1932).
SCHEFFER, J. D. R.: Untersuchungen über die Diffusion einiger organischer und anorganischer Verbindungen. Ber. dtsch. chem. Ges. **15**, 788—801 (1882).
— Untersuchungen über die Diffusion einiger organischer und anorganischer Verbindungen, II. Ebenda **16**, 1903—1917 (1883).
— Untersuchungen über die Diffusion wässeriger Lösungen. Z. physik. Chem. **2** 390—404 (1888).
— and F. E. C. SCHEFFER: On diffusion in solutions. Proc. Soc. Sci. Akad. Amsterd. **19**, 148—162 (1916).
SCHERP, H. W.: The diffusion coefficient of crystalline trypsin. J. gen. Physiol. **16**, 795—800 (1933).
SCHIØDT, E.: Swelling of erythrocytes in solutions of ammonium salts. Ebenda **16**, 977—986 (1933).
SCHUHMEISTER, J.: Untersuchungen über die Diffusion der Salzlösungen. Sitzgsber. Akad. Wiss. Wien, Math.-naturwiss. Kl. II **79**, 603—626 (1879).
SEDDIG, M.: Über die Messung der Temperaturabhängigkeit der BROWNschen Molekularbewegung. Physik. Z. **9**, 465—468 (1908).
SEITZ, W.: Über die Bestimmung des Diffusionskoeffizienten nach der elektrolytischen Methode von H. F. WEBER. Wiedemanns Ann., N. F. **64**, 759—777 (1898).
SEMEONOFF, E.: The diffusion of orthophosphate into and from muscle. Quart. J. exper. Physiol. **21**, 187—192 (1931).
SHAXBY, J. H.: Studies in Brownian movement. II. The determination of AVOGADRO's number from observations on bacteria (cocci). Proc. roy. Soc. Lond. A **104**, 655—667 (1923).
— Sur la diffusion de particules en suspension. C. r. Acad. Sci. Paris **180**, 195—197 (1925).
SHOUP, C. S.: The respiration of luminous bacteria and the effect of oxygen tension upon oxygen consumption. J. gen. Physiol. **13**, 27—45 (1929).
SIMMLER, TH. u. H. WILD: Über einige Methoden zur Bestimmung der bei der Diffusion einer Salzlösung in das reine Lösungsmittel auftretenden Konstante. Poggendorffs Ann. **100**, 217—235 (1857).
SITTE, E.: Über die „Fortpflanzungsgeschwindigkeit" der Diffusion und ihre Messung. Physik. Z. **32**, 410—414 (1931).
— Untersuchungen über Diffusion in Flüssigkeiten. V. Zur Theorie der Diffusion in Lösungen starker Elektrolyte. Z. Physik **79**, 320—344 (1932).
SMOLUCHOWSKI, M. v.: Zur kinetischen Theorie der BROWNschen Molekularbewegung und der Suspensionen. Ann. Physik (4) **21**, 756—780 (1906).
— Abhandlungen über die BROWNsche Bewegung u. verwandte Erscheinungen. OSTWALDs Klassiker der exakten Wissenschaften. Nr. 207. Leipzig 1923.
SOMERS, A.: Note on the attainment of a steady state when heat diffuses along a moving cylinder. Proc. physic. Soc. Lond. **25**, 74—76 (1912).

Stefan, J.: Über das Gleichgewicht und die Bewegung, insbesondere die Diffusion von Gasgemengen. Sitzgsber. Akad. Wiss. Wien, Math.-naturwiss. Kl. II **63**, 63—124 (1871).
— Über die Diffusion der Kohlensäure durch Wasser und Alkohol. Ebenda II **77**, 371—409 (1878a).
— Über die Diffusion der Flüssigkeiten. Ebenda II **78**, 957—975 (1878b).
— Über die Diffusion der Flüssigkeiten, II. Ebenda II **79**, 161—214 (1879).
Stella, G.: The concentration and diffusion of inorganic phosphate in living muscle. J. of Physiol. **66**, 19—31 (1928).
Stern, K. G.: Über die Diffusion des Hydroperoxyds in verschiedenen Lösungsmitteln. Ber. dtsch. chem. Ges. B **66**, 547—554 (1933).
Stewart, D. R.: The permeability of the *Arbacia* egg to non-electrolytes. Biol. Bull. Mar. biol. Labor. Wood's Hole **60**, 152—170 (1931).
— and M. H. Jacobs: The effect of fertilization on the permeability of the eggs of *Arbacia* and *Asterias* to ethylene glycol. J. cell. a. comp. Physiol. **1**, 83—92 (1932a).
— — The permeability of the egg of *Arbacia* to ethylene glycol at different temperatures. Ebenda **2**, 275—283 (1932b).
— — The effect of certain salt solutions on the permeability of the *Arbacia* egg. Biol. Bull. Mar. biol. Labor. Wood's Hole **65**, 368 (1933).
Stiles, W.: The penetration of electrolytes into gels. I. The penetration of sodium chloride into gels of agar-agar containing silver nitrate. Biochemic. J. **14**, 58—72 (1920).
— The penetration of electrolytes into gels. IV. The diffusion of sulphates in gels. Ebenda **15**, 629—635 (1921).
— The indicator method for the determination of coefficients of diffusion in gels with special reference to the diffusion of chlorides. Proc. roy. Soc. Lond. A **103**, 260—275 (1923).
— Permeability. London 1924.
— and G. S. Adair: The penetration of electrolytes into gels. III. The influence of the concentration of the gel on the coefficient of diffusion of sodium chloride. Biochemic. J. **15**, 620—628 (1921).
Süllmann, H.: Diffusionskoeffizienten und Teilchengrößen farbloser Stoffe (Zuckerarten, Harnstoff, Glyzerin, Urotropin). Protoplasma (Berl.) **13**, 546—566 (1931).
Sutherland, W.: First reported at a meeting of the Australian Association for the Advancement of Science 1904.
— A dynamical theory of diffusion for non-electrolytes and the molecular mass of albumin. Philosophic. Mag. (6) **9**, 781—785 (1905).
Svedberg, T.: Diffusionsgeschwindigkeit und Teilchengröße disperser Systeme. Z. physik. Chem. **67**, 105—111 (1909).
— Die Existenz der Moleküle. Leipzig 1912.
— Zentrifugierung, Diffusion und Sedimentationsgleichgewicht von Kolloiden und hochmolekularen Stoffen. Kolloid-Z. **36** (Zsigmondy-Festschrift), 53—64 (1925).
— Die Molekulargewichtsanalyse im Zentrifugalfeld. Kolloid-Z. **67**, 2—16 (1934a).
— Sedimentation of molecules in centrifugal fields. Chem. Rev. **14**, 1—15 (1934b).
— u. A. Andreen-Svedberg: Diffusionsgeschwindigkeit und relative Größe gelöster Moleküle. Z. physik. Chem. **76**, 145—155 (1911).
— and R. Fåhraeus: A new method for the determination of the molecular weight of the proteins. J. amer. chem. Soc. **48**, 430—438 (1926).
— and A. Hedenius: The sedimentation constants of the respiratory proteins. Biol. Bull. Mar. biol. Labor. Wood's Hole **66**, 191—223 (1934).

SVEDBERG, T. and J. B. NICHOLS: The application of the oil turbine type of ultracentrifuge to the study of the stability region of carbon monoxide-hemoglobin. J. amer. chem. Soc. **49**, 2920—2934 (1927).

— and H. RINDE: The ultra-centrifuge, a new instrument for the determination of size and distribution of particles in amicroscopic colloids. Ebenda **46**, 2677—2693 (1924).

— and B. SJÖGREN: The molecular weight of BENCE-JONES protein. Ebenda **51**, 3594—3605 (1929).

TAMMANN, G. u. V. JESSEN: Über die Diffusionskoeffizienten von Gasen in Wasser und ihre Temperaturabhängigkeit. Z. anorg. u. allg. Chem. **179**, 125—144 (1929).

TAYLOR, H. S.: A Treatise on Physical Chemistry. New York 1924.

TESCHENDORF, W.: Über die Resorptionszeit von Gasen in der Brusthöhle. Arch. f. exper. Path. **104**, 352—374 (1924).

THOULET, J.: Sur la diffusion de l'eau douce dans l'eau de mer. C. r. Acad. Sci. Paris **112**, 1068—1070 (1891).

THOVERT, J.: Sur une application nouvelle d'observations optiques à l'étude de la diffusion. Ebenda **133**, 1197—1199 (1901).

— Sur une application nouvelle des observations optiques à l'étude de la diffusion. Ebenda **134**, 594—596 (1902a).

— Diffusion retrograde des électrolytes. Ebenda **134**, 826—827 (1902b).

— Sur une conséquence de la théorie cinétique de la diffusion. Ebenda **135**, 579—580 (1902c).

— Recherches sur la Diffusion. Ann. Chim. et Phys. (7) **26**, 366—432 (1902d).

— Diffusionmètre. C. r. Acad. Soc. Paris **137**, 1249—1251 (1903).

— Relation entre la diffusion et la viscosité. Ebenda **138**, 481—482 (1904).

— Diffusion et théorie cinétique des solutions. Ebenda **150**, 270—272 (1910).

— Recherches sur la diffusion des solutions. Ann. Physique **2**, 369 bis 427 (1914).

TRAUBE, J. u. M. SHIKATA: Diffusion von Farbstoffen in Gele. Kolloid-Z. **32**, 313—316 (1923).

ULLMANN, E.: Experimentelle Beiträge zur Kenntnis der Diffusion in Lösungen. Z. Physik **41**, 301—317 (1927).

VANZETTI, B. L.: Untersuchungen über Diffusion von Elektrolyten und Niederschlagsbildung in Gallerte. Z. Elektrochem. **20**, 570—579 (1914).

VIERORDT, K.: Beschreibung eines verbesserten Endosmometers. Poggendorffs Ann. **73**, 519—531 (1848).

VOIGTLÄNDER, F.: Über die Diffusion in Agargallerte. Z. physik. Chem. **3**, 316—335 (1889).

VOIT, E.: Über die Diffusion von Flüssigkeiten. Poggendorffs Ann. **130**, 227—240, 393—423 (1867).

VRIES, H. DE: Sur la perméabilité du protoplasme des beteraves rouges. Arch. néerl. Physiol. **6**, 117—126 (1871).

WALLSTABE, FR.: Über die Diffusion von Radium-Emanation in Flüssigkeiten. Physik. Z. **4**, 721—722 (1903).

WALPOLE, G. S.: Counter diffusion in aqueous solution. Biochemic. J. **9**, 132—137 (1915).

WARBURG, O.: Versuche an überlebendem Carcinomgewebe (Methoden). Biochem. Z. **142**, 317—333 (1923).

WASHBURN, E. W.: An Introduction to the Principles of Physical Chemistry. New York 1921.

WEAVER, W.: The duration of the transient state in the settling of small particles. Phys. Rev. **27**, 499—503 (1926).

Weaver, W.: Die Diffusion kleiner Teilchen in einer Flüssigkeit. Z. Physik **43**, 296—298 (1927).

Weber, H. F.: Untersuchungen über das Elementargesetz der Hydrodiffusion. Wiedemanns Ann., N. F. **7**, 469—487, 536—552 (1879).

Westgren, A.: Bestimmung der Diffusion, der Fallgeschwindigkeit und des Sedimentationsgleichgewichts der Teilchen in Selen- und Goldhydrosolen. Z. physik. Chem. **89**, 63—90 (1914).

Wiedeburg, O.: Über die Hydrodiffusion. Wiedemanns Ann., N. F. **41**, 675—711 (1890).

— Über die Prüfung der Nernstschen Diffusionstheorie. Z. physik. Chem. **10**, 509—516 (1892).

— Zur Diffusionstheorie. Ebenda **30**, 586—592 (1899).

Wiener, O.: Darstellung gekrümmter Lichtstrahlen und Verwerthung derselben zur Untersuchung von Diffusion und Wärmeleitung. Wiedemanns Ann., N. F. **49**, 105—149 (1893).

Wilke, E. u. W. Strathmeyer: Experimentelle Beiträge zur Theorie der Diffusionsvorgänge. Z. Physik **40**, 309—321 (1926).

Williams, J. W. and L. C. Cady: Molecular diffusion in solution. Chem. Rev. **14**, 171—217 (1934).

Williamson, E. D. and L. H. Adams: Temperature distribution in solids during heating or cooling. Phys. Rev., II. s. **14**, 99—114 (1919).

Wollaston, W. H.: On double images caused by atmospheric refraction. Philos. trans. roy. Soc. Lond. **90**, 667—675 (1800).

Wright, C. I.: The diffusion of carbon dioxide in tissues. J. gen. Physiol. **17**, 657—676 (1934).

Wroblewski, S. v.: Über die Gesetze, nach welchen die Gase sich in flüssigen, festflüssigen und festen Körpern verbreiten. Wiedemanns Ann. N. F. **2**, 481—513 (1877).

— Über die Constante der Verbreitung der Kohlensäure im reinen Wasser. Ebenda **4**, 268—277 (1878).

— Über die Abhängigkeit der Constante der Verbreitung der Gase in einer Flüssigkeit von der Zähigkeit der letzteren. Ebenda **7**, 11—23 (1879a).

— Über die Natur der Absorption der Gase. Ebenda **8**, 29—52 (1879b).

— Über die Anwendung der Photometrie auf das Studium der Diffusionserscheinungen bei den Flüssigkeiten. Ebenda **13**, 606—623 (1881).

Wüstner, H.: Über Diffusion und Absorption von Wasserstoff in Quarzglas. Ann. Physik (4) **46**, 1095—1129 (1915).

Yégounow, M.: La diffusion des solutions et les poids moléculaires. C. r. Acad. Sci. Paris **142**, 954—957 (1906a).

— Diffusion des solutions de $CuSO_4$ dans la gélatine. Ebenda **143**, 882 bis 884 (1906b).

— Appareils pour diffusion dans les milieux solides. Arch. Sci. phys. et nat. (4) **25**, 350—359 (1908).

Zeile, K.: Über die Bestimmung von Diffusionskoeffizienten hochmolekularer Körper und über einige Beobachtungen bei der Diffusion von Katalase. Biochem. Z. **258**, 347—359 (1933).

Zuber, R.: Untersuchungen über Diffusion in Flüssigkeiten. II. Über einen Mikrodiffusionsapparat für ungefärbte Flüssigkeiten. Z. Physik **79**, 280 bis 290 (1932a).

— Untersuchungen über Diffusion in Flüssigkeiten. III. Diffusionsmessungen an elektrisch neutralen Flüssigkeitsgemischen und Lösungen. Ebenda **79**, 291—305 (1932b).

— u. K. Sitte: Untersuchungen über Diffusion in Flüssigkeiten. IV. Elektrolytlösungen. Ebenda **79**, 306—319 (1932).